Digital Journalism Practice

Ian Bucknell and Layla Painter

Digital Journalism Practice

Skills for now and the future

PETER LANG
New York · Berlin · Bruxelles · Chennai · Lausanne · Oxford

Bibliographic information published by the Deutsche Nationalbibliothek
The German National Library lists this publication in the German National Bibliography; detailed bibliographic data is available on the Internet at http://dnb.dnb.de.

Library of Congress Cataloging-in-Publication Data

Names: Bucknell, Ian author | Painter, Layla author
Title: Digital Journalism Practice : Skills for Now and the Future / Ian Bucknell and Layla Painter.
Description: New York: Peter Lang, [2026] | Includes bibliographical references and index.
Identifiers: LCCN 2026019924 (print) | LCCN 2026019925 (ebook) | ISBN 9781636671895
paperback | ISBN 9781636671901 ePDF | ISBN 9781636671918 epub
Subjects: LCSH: Online journalism | Digital media | Journalism--Technological innovations
Classification: LCC PN4784.O62 B83 2026 (print) | LCC PN4784.O62 (ebook)
LC record available at https://lccn.loc.gov/2026019924
LC ebook record available at https://lccn.loc.gov/2026019925

Cover image: istock/VPanteon

ISBN 978-1-63667-189-5 (Print)
ISBN 978-1-63667-190-1 (ePDF)
ISBN 978-1-63667-191-8 (ePUB)
DOI 10.3726/b23643

Published by Peter Lang Publishing Inc., New York, USA

info@peterlang.com

This publication has been peer reviewed.

www.peterlang.com

Contact for General Product Safety Regulation (GPSR): gpsr@peterlang.com

Contents

Exercises

Figures

Tables

Preface

This book was born of fear – a feeling that we know is shared by many teachers and students of journalism. Developments in the media feel at times overwhelmingly vast and quick. As new technologies and social media platforms emerge, each with their own practical, editorial and ethical challenges, it can feel impossible to keep up. Journalism students need teaching that will equip them with the skills demanded by this ever-changing industry, but it is not easy to deliver. One study of US journalism degree programmes in 2011 noted that the courses varied in their adoption of teaching multimedia skills 'due to budget constraints, varying loyalty to the media industry, and some professors' scepticism and fear of technology' (Sarachan, 2011, p. 160). This wariness of technology can be felt by teachers and students alike.

From our own experience of journalism and from conversations with people currently working in newsrooms, including seasoned editors and recent graduates, we can assure you that the wariness and fear is felt by those working in the industry as well. We hope in this book to alleviate some concerns by looking at overarching concepts in digital journalism, demystifying some of the technical processes and consolidating the need for and use of traditional journalism skills on new platforms.

We unfortunately don't have a crystal ball and so we can't predict exactly how journalism will look in years to come. Even in the time it's taken to write this book we've seen social media platforms rise and fall, leaps in technology and changes in trends. The pace and scale of development can be daunting but it also brings opportunity. As a digital journalist you will find occasions when you are the first to try a new format. You can become the expert in a new technology if you take the time to study how and why it can work for you and your journalism. We hope to equip you with the confidence to overcome the fear and to apply your journalism skills in innovative ways – whatever the future holds.

Acknowledgements

This book relies heavily on reflecting the experience of those digital journalists and experts who have shared their knowledge with us. We'd like to thank all those who have taken the time to speak with us, including graduates from our BA Journalism at the University of Leeds, now working in creative digital roles in the industry: Jacob Tomlinson (BBC); Georgia Carroll (Channel 4); Ryan Daniels (CBS Sports); Jack Gilmartin (Channel 4) and Laura Walcott (digital marketing strategist). Dr Rebecca Whittington (Reach Plc) generously shared her expertise, based on academic study and industry practice, in keeping journalists safe when working online. We also gained insights from a range of digital editors from varied news organisations around the UK. They shared their thoughts anonymously from the perspective of senior journalists involved in shaping digital strategy, as well as managing daily production. Thank you to them and their employers at Sky News, ITN, the Financial Times, BBC, Vice and JPI Media.

We would also like to acknowledge the support of colleagues from the School of Media and Communication at the University of Leeds. Professor Julie Firmstone has mentored us through the professional transformation from journalists to academics capable of scholarly research and writing. Julie also provided feedback on the draft manuscript, as did our colleague Carl Hartley, drawing on his expertise in audio journalism. We had help with photography, and in some cases modelling, from Samson Borthwick, Izzy Brittle and Charles Balatoni, a group of people who quietly keep our School going and make work life enjoyable. Finally, we'd like to acknowledge the help of Professor Katy Parry, the Director of Research in our School. Katy has found new ways of working that encourage and enable academic colleagues from industry backgrounds, like us, to carry out research and share our findings through publication.

Introduction

This book is designed to help you gain the skills and understanding needed to become a digital journalist. It complements and builds on your studies of journalism and the media, taking many of the fundamentals you are learning and showing you how they play out in a digital environment. Learning a broad subject that is constantly changing is challenging, so our text is designed to take away the strain by focusing on key areas that are vital to digital journalism and, crucially, will continue to be of great importance in the future. We have done this by taking a 'five-dimensional approach' – focusing on practice, skills, audience data, safety and future proofing.

In Chapters 1–4, we focus on the *first dimension*, production practices and storytelling formats that have taken hold since digital media transformed news reporting. These include the social short videos that you'll frequently see in your feeds, online news reporting with multimedia components, digital audio journalism (such as podcasting) and the creative blending of text and images to tell stories using the tools available in social media. Alongside this, we will show you the new and old skills (*dimension 2*) needed to create digital journalism, no matter the media or format. These are covered in Chapter 6 and include filming using your smartphone, verifying information, finding stories and conducting interviews.

But it doesn't stop at knowing how to produce reports for digital media. Contemporary journalists also need a handle on what their audience wants and how to reach that audience through third-party platforms, over which they have no control. This understanding can be gained through analysing audience data (*dimension 3*), a topic we focus on in Chapter 5. You will learn how social media differ in character, the demographics of the users they attract, how people interact with posts and the impact analytics have on which stories are chosen and how they are told.

The benefits of everyone being able to engage with digital media are, alas, offset by those who use the platforms to try and cause harm. Journalists are among the professionals who come under attack from people, or malicious robotic accounts, that abuse and make threats. Alongside this, is the risk to our well-being of exposure to distressing content that is freely and frequently shared, and often of importance to journalists. The good news is that there are steps you can take and ongoing practices you can adopt that will protect you and others. You will learn about them in Chapter 7 that covers working safely online (*dimension 4*). As you are guided through the risks and the countermeasures, you will produce your own personal online safety plan that you can carry into your careers as media professionals.

Phew! That got a bit heavy. But we end on a high with a look at what else you can do to future proof (*dimension 5*) yourselves as journalists working with evolving computational technologies. Excitingly, you have the potential and opportunity to keep up with the curve by getting ahead of it, through combining your native knowledge of digital media, the apps you have grown up using, with your growing knowledge of journalism practice. You can become a digital journalism innovator and a model explaining one approach to achieving this is explained in Chapter 8. In the same chapter, you'll learn about the impact of artificial intelligence on newsrooms and the soft skills you'll need to keep thriving as a digital journalist.

As you'll see, our teaching is informed by academic studies, industry research, analysis of journalism production and our experience as teachers and practitioners of journalism. We also draw from our students' experience of learning at the University of Leeds. Many of our graduates are making successful careers as digital journalists and they are among the industry specialists who have contributed to this book, testifying to what students of the subject need to learn and offering advice on how you best to learn it.

We are driven by the knowledge that digital journalism is the growth area for jobs in the industry. It is therefore vital that you leave academia with the requisite skills, experiences and knowledge to compete for entry level work in the field. It would be impossible for you to learn every aspect of digital journalism production in evidence throughout the world, so this book uses industry and academic evidence to identify and explore the most important practices, skills and knowledge areas that will ensure your education remains relevant to your future careers. By learning how to create and disseminate stories using the approaches we cover, you will not only be ready to carry out the tasks that publishers and broadcasters want their employees to perform now, but will gain an understanding of the principles of digital journalism production that can be applied to formats, platforms and media that have not yet been developed.

In the remainder of this introduction, we look at how 'digital journalism' is defined and how it's evolved in recent decades, from the beginnings of the internet to the emergence of generative artificial intelligence. Special attention is paid to how social media has revolutionised the work of journalists, as we gauge how roles are changing as new media surpasses old. Finally, you will learn how to navigate the book, following the exercises that will see you put your newfound knowledge into practice and provide you with further opportunities to learn from experience.

Definition of 'digital journalism'

Digital journalism was defined in the early years of this century as 'the use of digital technologies to research, produce, and deliver (or make accessible) news and information to an increasingly computer-literate audience' (Kawamoto, 2003, p. 4). It was acknowledged that this definition was a 'moving target', as changes in technology and the conceptualisation of journalism would lead to changes in how we understand the term. A study by Columbia University into the commercial challenges associated with digital journalism showed this to be the case, as they asserted that hand-held devices and tablets should be included in the definition, to make clear that we are not limiting ourselves to journalism published online and viewed via a desktop computer (Grueskin et al., 2011, p. 4). This seems like a worthwhile clarification given that in the UK, mobile phones surpassed computers as the primary device for accessing digital news back in 2017 (Newman et al., 2019, p. 68).

Any contemporary definition of digital journalism must also acknowledge the significance of social media as a major gateway to news. In the USA, it's reached a point where 48 % of the population regularly use social media as a news source (Newman et al., 2024, p. 115), whilst in the UK the figure is 37 % (ibid. p. 65). Dependence on social media is higher amongst younger people and still growing, with 43 % of 18- – 24-year-olds around the world saying that it's their main news source, up from 18 % in 2015 (Newman et al., 2023, p. 12). This is attributed to the growing number of 'social natives' within this age group. As you'll discover, creating news stories that suit social platforms and appeal to their growing number of users, is a key part of being a digital journalist.

So, to summarise, the definition of 'digital journalism' has evolved from using computers to make news that is viewed on computers, to also covering the production of stories for smartphone apps and social media. This theoretical definition of the practices covered by the term 'digital journalism' is aligned with its application in the industry. Digital teams within news organisations worldwide are typically responsible for managing online and social media content creation and distribution, aware that their work will typically be viewed on a mobile device.

Digital journalism is not about the tools reporters use to gather their stories, but more about what they do with the material afterwards. For instance, using a smartphone to capture images or audio for television or radio broadcasts, while involving digital technology in the recording process, does not fall under our definition of digital journalism, as the final product is distributed through traditional media channels. Conversely, utilising conventional video cameras or audio recorders to produce news content specifically designed for websites, mobile apps, and social media platforms, does constitute digital journalism. This book aims to equip you with the necessary skills and knowledge to plan, create and share such journalism, helping you build your professional profile and increasing your employability in the digital departments of media organisations.

Evolution of digital journalism

The digital journalism revolution has seen vast changes in just a few decades. We'll provide some historical context throughout the book. Here is a brief timeline.

Table 0.1: A brief history of digital journalism

1965	First computers are networked	The ARPANET linked computers so they could share information. This was the beginnings of the Internet.
1970	Teletext invented	BBC Ceefax launched in 1971 and discontinued in 2012.
1971	First email sent	By US computer programmer Ray Tomlinson, to himself.
1974	Videotext invented	
1978	Bulletin Board Systems	These were the first systems to allow computer users to log in to a network and exchange data.
1980	First online newspaper	The Columbus Dispatch in July 1980, part of a pilot between dial-up service CompuServe and Associate Press.
1990	World Wide Web invented	Tim Berners-Lee, CERN develops FTP (File Transfer Protocol) as a method of moving and storing data, and HTML (Hypertext Mark Up Language) – the code that allows information to be displayed on 'web pages'.
1993	First web browser	Mosaic allows people to search and navigate the World Wide Web.
1995	CNN news website launches	

1997	First social media site launches BBC News online launches	SixDegrees.com is widely seen as the first social media website, although elements of social networking were available before this date.
2003	Myspace launches	Biggest social networking site up to this to date.
2003	First podcast	IT Conversations by Doug Kayne.
2004	Facebook launches	Still the biggest social network worldwide.
2005	YouTube launches Huffington Post launches	First video streaming platform.
2006	Twitter launches Buzzfeed launches	Twitter posts are limited to 140 characters and so introduces 'microblogging' – telling stories in short form.
2007	First iPhone launches	The smartphone revolution really begins.
2010	Instagram launches	
2011	Twitch launches	First public live streaming platform.
2017	TikTok launches	
2020	Clubhouse launches	First specifically audio social media platform.
2022	ChatGPT released	Generative AI becomes easily accessible.
2024	Bluesky launches	Alternative microblogging platform opens for registration in February 2024, with over 26 million users by January 2025.

As *table 0.1* illustrates, there have been many milestones on the road to the digitisation of journalism, with legacy news organisations changing their practices along the way and the emergence of digital-native publishers. This has transformed the journalism labour market. The situation is stark in the United States where, according to a report by the Pew Research Center, 'newspaper newsroom employment fell 57 % between 2008 and 2020, from roughly 71,000 jobs to about 31,000. At the same time, the number of digital-native newsroom employees rose 144 %, from 7,400 workers in 2008 to about 18,000 in 2020' (Walker, 2021). The same pattern is seen throughout the world, for example in Spain where a study reports that there were 1,382 active digital-native news websites in Spain by 2019, a more than tenfold increase from the 116 counted in 2005 (Negredo, 2020, p. 76).

The impact that social media has had on the work of journalists is also clearly established. It is some years now since a survey of around 700 UK journalists by the Reuters Institute for the Study of Journalism, found that the biggest change experienced by reporters related to the impact of social media. 98 % of respondents said they believed that the influence of social networks had increased in the previous five years (Thurman et al., 2016, p. 36). These platforms have changed the way that journalists work, as many now use social media as channels for delivering news, as well as researching stories and interacting with audiences. In some instances, this is a case of reporters in non-digital roles adding new approaches to the ways they have always worked. But the evolution of digital journalism has also led to the rise of social media journalists; discrete roles that sit alongside legacy positions:

> Social media teams are increasingly viewed as (1) the first stage in the customer acquisition funnel and (2) a means of creating a culture that puts audience impact and digital optimisation first. Their role has grown from simply ensuring content is optimised for specific platforms to originating content tailored for those destinations. (Kueng, 2017, p. 26)

Preparing future digital journalists

The birth of the social media reporter is one example of how digital media has revolutionised journalism and the role of reporters. This has clear implications for aspiring journalists and those who shape the minds of future newsmakers. Jonathan Baker, the former Director of the BBC College of Journalism, agrees. In his provocative essay 'Get digital or die', he calls on educators to furnish journalism students with skills relevant to the digital world, or risk their programmes becoming obsolete (Baker, 2016). This truth is one of the driving forces behind our five-dimensional approach to teaching digital journalism practice.

Whilst there is agreement over the need for journalism education programmes to embrace digital practices, there is little guidance available on how this should be done, perhaps due to the scale of the challenge. Piet Bakker (2014) identified four growing areas of digital journalism activity (technical, community, content and commercial) in an article on how the skills journalists need and the tasks they perform have fundamentally changed. In his conclusion, he acknowledges the difficulties created for journalism educators by 'the diversity and the depth of the skills requested. It is impossible for any journalist to master them all, let alone for any J-school to teach all these skills in depth' (Bakker, 2014, p. 603).

Furthermore, news providers themselves have divergent and ever-evolving digital strategies, often forced upon them by the dominance of third-party platforms. In 2018, the New York Times Chief Executive Mark Thompson made plain his frustration with the power wielded by social media companies:

> We do not know, beyond inevitably imperfect and incomplete empirical observation, how the algorithms of the major platforms sort and prioritise our content, nor can we reliably predict or influence changes in those algorithms, nor in any sense hold the companies to account for them. (cited in Mayhew, 2018)

Along with a lack of control over digital distribution, news organisations must struggle with the related issue of the rapid pace of innovation and change in digital media. Keung describes how 'shiny new things' have at times tempted publishers and broadcasters into decisions they later regret: 'Strategies are being thrown off course by what feels like a ceaseless stream of new technologies and products. These can be steppingstones to the future, or expensive distractions from the core mission' (Kueng, 2017, p. 21). If it is unwise for industry to be over-eager in the pursuit of currency, the same must apply to educational institutions.

So, there are seemingly two strands of research that are difficult to reconcile: one that stresses the urgent need for journalism educators to teach the digital skills and practices required by industry, the other that has found evidence suggesting that industry is still working out what those skills and practices should be. How do educators serve the needs of industry, when industry itself is yet to define what those needs are? This book seeks to resolve this dilemma.

We set about doing this by focusing on practices highlighted in a study of digital production amongst UK news providers that drew on the insights of digital editors from a cross-section of the news media (Bucknell, 2020). The participating organisations included vice.com (UK), the Financial Times, Sky News and the BBC, thereby representing commercial and public service broadcasters, digital natives and newspapers. Digital editors were selected for interview because they are ideally placed to understand the strategies and aspirations of their employers, as well as having daily, first-hand experience of how digital journalism is produced on the newsroom floor. During semi-structured interviews, they identified a range of practices, formats, skills and knowledge, some of which were particular to their own organisations, but many that recurred across the sample. This suggested a degree of consistency and conformity had emerged in digital journalism production that is helpful to those teaching and learning the subject. From these findings, along with further interviews with industry experts and analysis of the digital content created by news organisations, we have crafted our five-dimensional strategy.

The active audience

Another important theme in this book, reflecting a central driving force behind decisions and change in digital journalism, is the fundamental difference in the relationship between news organisations and audiences in an online environment. Much has been written about the democratisation of the media – the shift of power from organisations to individuals. Members of the public have easy access to the same technology to create content and share from their own phones. Huge stories can be broken by passers-by or amateur journalists before the professional journalists even know it's happening. YouTubers and influencers can hold as much sway as correspondents.

In addition, the previous one-way communication between the media and readers has become a two-way process. This is what John Naughton (2012) refers to as the change from a 'push' medium (broadcasters create content and push via broadcast channels towards passive recipients) to a 'pull' medium (where individual recipients can choose what they want to see or read and pull it down to their computers). Readers can also easily share opinion on stories – comment on and discuss articles, vote in online polls and react with immediacy to something they've watched or read. Journalists can also use the internet to find stories, monitor what is happening, source case studies and user-generated material.

The online news environment has not only changed how users interact with the news, but it has also provided journalists with more detail than ever before about their audiences (Napoli, 2011). Web analytics allow news organisations to find out a vast amount of information – where and who their audiences are, what and when they view content and how long for, what they are interested in and they chose to interact with. Studies have shown that this information guides subsequent editorial decisions. For example, online newsroom editors have been shown to rely on this data to understand the popularity of news items to promote content that audiences will be more likely to click on. (Lee et al., 2012) This raises questions over the extent to which this would sway news organisations towards stories that the public are interested in, rather than those in the public interest. Or whether – in the race for a higher number of clicks, content is being produced to fit the priorities of a particular platform, rather than the editorial values of a news organisation.

However, most journalists and editors maintain that editorial decisions are still mostly based on traditional news values, rather than whether stories would achieve significant numbers of clicks (Hanush, 2017). This underlines the importance of continuing to emphasise these news values in the journalism classroom. Throughout, this book will encourage you as a digital journalist to demonstrate this news sense in choosing what stories to cover, whilst also having an awareness of the analytical tools available and how to interpret that data to tell and distribute your story effectively (Chapter 5).

Critical context

In the Digital Journalism module that forms part of the undergraduate degree we teach at the University of Leeds, we run a lecture series alongside the practical workshops. This creates space for students to reflect on how the role of journalism is being transformed by new technologies and how their experience of reporting for digital services connects to these changes. Our philosophy is that there are reciprocal benefits to the study and practice of journalism; the more you understand the subject, the better the journalist you will become, and the more you practice journalism, the better able you are to study the subject.

Although wrestling with the many and complex issues associated with digital journalism production is not the purpose of this book, we have drawn on our philosophy to highlight, in places, the critical context for the practices you will learn. Attention to relevant research by academics and industry bodies is made throughout the book and you will find references to our sources, plus further recommended reading, at the end of each chapter. We encourage the curious journalism student to seek out these sources and broaden your knowledge. There are points where the issues related to the topics we cover are so pressing that we do pay them further attention. This can be seen in subchapters such as the *critical issues* addressed in Chapter 5, where concerns about the potential over-reliance on audience analytics to guide editorial decisions are discussed, or in Chapter 8, where we consider how ethics and standards are impacted by using artificial intelligence in the production of journalism.

How to use this book

Reading this book from front to back will give you a comprehensive understanding of what digital journalism is and how you do it. But the chapters also work discretely so, if you want to jump straight in with 'Digital audio', go ahead. You'll also notice that we cross-reference between the inter-linked formats, skills and issues covered. This allows you, for instance, to look up how you film using a smartphone during the section on social short video production, if you're not already comfortable with using your mobile to shoot.

An important feature of our approach is that we regularly set you exercises that prompt you to put into practice the skills, routines and thought processes that we cover. This allows you to instil ideas while they are fresh in your minds, giving you confidence that you have understood the concepts and processes. Experiential learning is vital to becoming a media professional, so these exercises not only highlight what you've already gleaned, but in themselves provide you with further opportunities to learn and we'll prompt you along the way to reflect on your practice to maximise what you can gain. Your tasks are clearly highlighted in breakout boxes, illustrated by *Exercise 0.1*.

Exercise 0.1: The tasks we want you to carry out are spotlighted using boxes

Within each chapter of this book, we have suggested a series of exercises. Some are critical and concern the analysis of current digital journalism, but most are practical and put the skills and concepts we've discussed to use.

The exercises have been designed to work as independent learning opportunities for you, or to be used in a classroom by your tutors. These exercises are clearly marked in the relevant sections.

The headings and subheadings are there to help you find your way around the book. They vary depending on the content of each chapter but always provide a structure that will enable you to navigate your way to becoming a digital journalist. The production Chapters 1–4 start by familiarising you with the formats that we are focusing on, analysing their component parts, followed by guidance on how to produce reports of that type yourselves. We also consider what you need to do to promote your work, so that it reaches an audience and has impact.

References and recommended reading

Baker, J. 2016. Get digital or die: News storytelling, social media and journalism education. *Journalism Education*. **5** (2), pp. 9–11.

Bakker, P. 2014. Mr Gates Returns. *Journalism Studies*. **15** (5), pp. 596–606.

Boczkowski, P. and Anderson, C. W. 2017. *Remaking the News: Essays on the Future of Journalism Scholarship in the Digital Age*. Cambridge, Massachusetts: MIT Press.

Bucknell I. 2020. Going digital, not dying out: how universities are uniquely placed to teach digital journalism survival skills. *Journalism Education*. **9** (1), pp. 6–17.

Grueskin, B., Seave, A. and Graves, L. 2011. *The Story So Far: What We Know About the Business of Digital Journalism*. New York: Columbia University Press.

Hanusch, F. 2017. Web analytics and the functional differentiation of journalism cultures: individual, organisational and platform-specific influences on newswork. *Information, Communication & Society*. 20 (10), pp. 1571–1586.

Kawamoto, Kevin ed. 2003. *Digital journalism: emerging media and the changing horizons of journalism*. Oxford: Rowman & Littlefield Publishers.

Kueng, L. 2017. *Going Digital: A Roadmap for Organisational Transformation*. Oxford: Reuters Institute for the Study of Journalism.

Lee, A. M., Lewis, S. C. and Powers, M. 2014. Audience Clicks and News Placement: A Study of Time-Lagged Influence in Online Journalism. *Communication Research*. **41** (4), pp. 505–530.

Mayhew, F. 2018. New York Times boss Mark Thompson: 'When it comes to news, Facebook still doesn't get it'. *Press Gazette*. [Online]. [Accessed

8 July 2022]. Available from: <https://pressgazette.co.uk/new-york-times-boss-mark-thompson-when-it-comes-to-news-facebook-still-doesnt-get-it/>

Napoli, P. M. 2011. *Audience Evolution – New Technologies and the Transformation of Media Audiences*. New York: Columbia University Press.

Naughton, J. 2012. *From Gutenberg to Zuckerberg: What You Really Need to Know About the Internet*. London: Quercus.

Negredo, S., Martínez-Costa, M-P., Breiner, J., and Salaverría, R. 2020. Journalism Expands in Spite of the Crisis: Digital-Native News Media in Spain. *Media and Communication*. **8** (2), pp. 73–85.

Newman, N. R, Fletcher, R., Eddy, K., Robertson, C-T and Nielsen, R. K. 2023. *Reuters Institute Digital News Report 2023.* Oxford: Reuters Institute for the Study of Journalism.

Newman, N. R, Fletcher, R., Robertson, C-T, Arguedas, A. R. and Nielsen, R. K. 2024. *Reuters Institute Digital News Report 2024.* Oxford: Reuters Institute for the Study of Journalism.

Thurman, N., Cornia, A. and Kunert, J. 2016. *Journalists in the UK*. Oxford: Reuters Institute for the Study of Journalism.

Walker, M. 2021. U.S. newsroom employment has fallen 26 % since 2008. *Pew Research Center*. [Online]. [Accessed 8 July 2022]. Available from: <https://www.pewresearch.org/fact-tank/2021/07/13/u-s-newsroom-employment-has-fallen-26-since-2008/>

5 July 2021. [illegible] bo-miami-[illegible]
Nordl, P. M. 20[illegible]
thestylefa[illegible]
Naughton, J. 20[illegible]
Lone Abro[illegible] (Q[illegible]
Negredo, S. Martin[illegible] M. R. Bonnet[illegible]
Journalism [illegible] the Crisis: Dig[illegible]
Spain. Media [illegible] 8 [illegible] pp. 73[illegible]
Newman, N. R. Fletcher [illegible] Robertson [illegible]
Reuters Institute Digital News Report [illegible]
Study of Journalism.
Newman, N. R. Fletcher [illegible] Robertson, [illegible]
R. K. 2024 Reuters Institute Digital News Report 2024. [illegible]
Institute for the Study of Journalism.
Thurman, N., Cornia [illegible] Kunert, J. 2016 [illegible]
Reuters Institute for the Study of Journalism.
Walker, M. 2021. U.S. [illegible] has fallen [illegible]
Research Center [illegible] 2021 [illegible]
www.pewresearch.org [illegible]
has-fallen-26-since-[illegible]

CHAPTER 1

Digital video

We've chosen to start the book with digital video production because of its importance as a means of distributing journalism on social media, mobile apps and the web. This goes against the convention of textbooks on our subject that usually begin with online news production, an approach that makes sense if you've lived through the evolution of digital journalism, starting with the migration from print to the internet. But that means little to you 'social natives' who are currently studying journalism in a world where there is growing demand for video news on digital platforms. The Reuters Institute reported that 66 % of their global survey respondents said they watch short online news videos every week (Newman et al. 2024, p. 13), so starting with where digital journalism is now, rather than where it's been, is our priority.

In this chapter, we'll look at different types of news video created for digital distribution. The focus is 'social shorts' due to how prevalent they are in industry around the globe and their convenience as a way of learning key principles. They are short form pieces that you can create quite quickly whilst demonstrating skills and understanding that could be applied to longer forms. You'll also learn about video explainers, a close relation of the social short that is also designed to meet the needs and behaviours of a digital audience. You'll get to grips with the editorial values, story structures and production practices relevant to these formats, reaching your own conclusions through exercises that guide your analysis and reflections. Ultimately, you'll produce your own digital videos, experiencing first-hand what it takes to create and disseminate engaging stories that both suit the appetites of social audiences and uphold journalistic standards.

Social short videos

What is a social short?

Social short videos are created specifically for social media distribution. They look different to traditional TV news packages as they are presented in the vertical orientation to match the screens of smartphones, 'the most important digital device for accessing news across countries' (Newman et al., 2022, p. 22). Social shorts include captions that both tell the story and transcribe the words of interviewees. You can see what this looks like on screen in *figures 1.1 and 1.2*. The BBC and Washington Post videos are designed with mobile viewing in mind with an aspect ratio (width by height proportion) of 9:16, the same dimensions as your smartphone. They look similar, despite being produced by two contrasting news organisations; the BBC is a UK broadcasting corporation, while the Washington Post is a USA newspaper in origin. Whoever the publisher is, there are characteristics of the social short format that are common around the world and across the media.

Figure 1.1: Screen grab of social short from BBC Yorkshire posted on Facebook.

Figure 1.2: Screen grab of social short the Washington Post on Instagram.

Alongside moving images shot by the publisher, they often incorporate UGC (user-generated content i.e. media created and shared by the public), graphics and still images. These videos are 'gaining momentum because they seem to fit with social media's structural designs and viewing patterns' (García-Perdomo, 2021, p. 8). Their popularity is widespread, from French digital news startup Brut, who specialise in the format, to traditional heavyweights like the New York Times producing shorts for Instagram. A digital editor from Sky News describes how people's behaviour on social networks influences the duration of social shorts: 'they don't consume for as long a period of time, so when a typical TV VT might be three to four minutes, a social short video will be 60 to 90 seconds'.

Unhelpfully, more than one term has emerged from industry and the academy to describe them and we've seen them referred to as digital shorts (as opposed to analogue underpants), digital videos, online videos and many other combinations of these same words. We'll refer to them as social shorts in this book.

Exercise 1.1: Analyse social shorts to understand their characteristics

The best way to get a handle on what social short videos are and how they're produced is to watch and analyse examples from a range of publishers. Have a look at the work of innovators such as NowThis, the USA-based pioneers of social video, or the shorts produced by Spanish digital native elDiarioes, available on their YouTube and Instagram channels. Also consider the work of legacy news organisations: Facebook videos from BBC News, the Washington Post's YouTube shorts or Al Jazeera's AJ+ website. Think about the following qualities and how they differ to other types of video output, such as traditional TV news:

- The editorial thinking behind the video. What story has been chosen and why might it suit a social audience? Who features in the video and why?
- Consider the structure of the films and in particular, the images and words that are used at the beginning. What does the narrative arc look like in a social short?
- Think about the captions, in terms of both production and storytelling. Are they the same throughout or do they differ for a purpose?

Story selection

The first discussion point above helps to identify what types of stories are chosen for social short production. For decades, journalism scholars have wrestled with the question of how journalists choose stories and the concept of 'news values' has come to the fore as a framework for categorising those decisions. Tony Harcup has investigated the evolution of news values and what they tell us about how stories are selected and covered (2015, p. 37–52). The way in which news organisations seek to understand their audience is a recurrent factor and in a digital media environment, that is calculated with precision. Story choices are often borne out of metrics (more of that in Chapter 5 on *Understanding audience data*) with digital producers searching to see which stories are getting engagement on their own channels or by looking at what topics are trending across the internet.

A universal conclusion that social video creators have reached through their study of analytics, is that popularity is less about the subject matter and more about the capacity for the story to trigger feelings. They've discovered 'stoking people's emotions is one of the best methods news producers have to increase distribution and engagement' (García-Perdomo, 2021, p. 10). Jacob Tomlinson, a social media journalist from the BBC agrees, explaining how he looks for 'stories that make your face move'. If people react emotionally, they are more likely to share the social short, increasing its reach and impact.

So, stories that have the capacity to emotionally effect the audience are prioritised. Inspirational tales of success, heroism or triumph over adversity are often featured, along with personal tragedies. In other words, the kind of stories that are shared on social media outside of journalism.

Social media journalists have found that another key attribute is stories with strong images that will grab the viewers' attention in an endless feed of posts that is being scrolled through at pace. In some cases, social shorts have little appeal beyond the initial hit of a captivating picture, explaining why low-hanging fruit, such as shots of aeroplanes trying to land on a windy day, are often shared by news organisations. In *figure 1.1*, we see a screenshot from a story about a boy with a disability that has many of the social story values we've alluded to so far. The joy on the boy's face provides a powerful image and the story is moving and inspirational, though it's borne of tragedy.

There is also evidence of topics being chosen for social shorts that would likely appeal to younger audiences who receive a high proportion of their news through social networks, such as coverage of environmental or educational issues, social injustices and celebrity news (Newman et al., 2022, p. 44). In his study of social shorts in Colombia, García-Perdomo identifies the prominence of quirky stories, as 'producers have realised that online audiences tend to share and react to unexpected, dramatic, odd, and sensational videos' (2021, p. 14). In each case, editors and producers are called upon to think about not only the newsworthiness of the story or event, but how to make it connect with a social audience.

Summary

Look out for these qualities when choosing a story for a social short video: trending topic; emotional hook; strong images; subject that appeals to younger audiences and quirkiness/humour.

Storytelling

As well as choosing which stories are told, how stories are told is also important in the creation of social short videos and again, the affordances of social networks and the behaviour of users is paramount. In line with the need to make an emotional connection to the audience, social short news stories are often told through the experience of a citizen source, usually a case study. The story of poverty caused by high rents, for example, can be told through the personal experience of a family that is struggling to cope, rather than statistics, expert insight and political reaction, as would be the case in a TV news piece on the same subject.

Directly conveying the experiences of those involved, rather than using third parties, was advocated as a virtuous news value long before social media came into being (Harcup, p. 41). But in the context of digital production, it is valued not only because of editorial merit but also to get in sync with how people behave on social platforms. As the Reuters Institute says of Facebook, 'people are mainly there to connect with friends and family rather than consume news' (Neman et al., 2022, p. 26). Producers align themselves with those motivations by presenting the stories of individuals who are relatable to the audience, i.e. other citizens.

The Financial Times, which has amassed over a million digital subscribers, says that their best performing social videos are 'on one specific story or one specific topic'. That fits with the relatively tight duration of the format: rather than salami slicing lots of angles into your 60–90 second film, it's best to focus on one viewpoint and do it justice.

The need to grab users' attention quickly in the constant flow of social media posts is an established priority across the globe. *Figure 1.1* illustrates a video that hooks us in by starting with the key, life-changing moment for the boy with cerebral palsy and his family, emphasised by the caption that underlines the significance of the moment. Ashante Bean, writing for the Poynter Institute in the USA, has produced a guide for publishers to creating and disseminating social video, called 'VidSpark':

> During the first few seconds of watching a video on social, the viewer is deciding whether they'll continue watching the video or keep scrolling. This makes the first seconds critical in creating a hook and delivering valuable content, giving the viewer a reason to continue watching. Skip the exposition, put the most interesting visuals and information upfront, before any branding. (Bean, 2021)

A digital editor in the UK agrees with this approach and takes it further, describing how at Sky News they invert the storytelling so that you don't get the reporter leading like you do in a TV piece, you get the real human element of the story straight at the top'. As previously discussed, appealing to our emotions is a key news value in social short videos, so it is no surprise that the human angle is prioritised in this way. It is often achieved by placing an emotive quote at the beginning of a video, as in *figure 1.1*.

Other methods of grabbing the audience's attention straight away include leading with dramatic pictures or choosing stories that are naturally visual, such as a weather event (see *figure 1.2*), festival or CCTV footage of a crime. Front loading with hard-hitting facts is another way to draw in your social media audience and one that can be combined with strong images. If people watch a video for more than three seconds on Facebook, that counts as a view and therefore starting with strong material helps hit that initial target.

But social media algorithms often reward content makers who hold our attention for longer. Jacob from the BBC says his team has noticed that videos that keep lots of viewers for around 20 seconds do well on Facebook, indicating that the platform has given them a prominent place in their users' feeds. So, after making an impactful start, the next challenge is to retain the attention of the viewer with further hard-hitting quotes, facts and images. In this short form, strip it down to the most important and striking details as there is evidence that allowing the pace to drop leads to a big fall in retention. Jacob cites an example where the data showed 50,000 viewers stopped watching a particular social short during a three second sequence where pictures of a forest had been allowed to breathe. He describes it as a 'wake up call' to the team, alerting them to how quickly audiences can lose interest.

Exposition (background and contextual information) is still a part of the storytelling in social short videos, but it follows the action rather than leading into it. Conclusions are kept short and, in keeping with TV news packaging, they are often forward-looking, referencing a future milestone in the development of the story. The overall effect is a disruption to the traditional TV narrative arc, where reporters first set the scene. In social shorts, the audience is thrown into the midst of the action with the most hard-hitting material and testimonies at the top, creating the following structure:

Action – > Exposition – > Further action – > Conclusion.

Tomlinson describes this social short narrative arc as being shaped like a slide rather than a hill, with the best material first, followed by the second best, third best and so on.

Summary

Good storytelling with social short video includes: a compelling case study; focus on one angle; grabbing audience's attention quickly; keeping up the pace to maintain that hold; inverting the narrative structure so that exposition (context and background) follows the main action.

Production

We now know what social shorts are and the thinking behind story choices and structure. But how do you create this type of video? A good starting point is to draw from the copious examples provided every day by news organisations around the world. There is no single approach but there are common features that we can identify and use as building blocks for our own work.

Filming

You'll learn mobile filming skills in Chapter 6, but it's worth noting now that social short producers say they tend not to film 'in-app' (i.e. using the camera function within a social media application) because it does not fit with their editorial workflow, where an editor needs to review the piece before it is published. Avoiding working in-app also reduces the technological load on reporters, allowing them to create their story using the video editing software they are familiar with before uploading it to all suitable social channels. You can add to your own video footage with photos, user-generated content and graphics, ensuring that you receive permission to use material that is not your own or that the images are freely available under a Creative Commons licence.

Audio

Although there is evidence that many viewers of social video do not listen to the audio (hence the inclusion of captions), those who unmute expect good sound quality, so do not skimp on the effort you put into recording clean audio with the correct microphone and put as much care into your audio edit as you would for TV. To accompany the segments where captions drive the narrative, include an appropriate piece of instrumental music. There are lots of audio databases that provide production music that is free to use for academic and other non-profit purposes. Have a look at the library offered by YouTube or websites such as Uppbeat and Pixabay.

Editing

Once you've gathered your audio and video, it's time to start editing. The pre-eminence of video editing software changes over time but all the major players that are familiar to TV producers (such as Adobe Premier Pro, Apple's Final Cut Pro and Avid Media Composer) provide solutions for editing social video, as do mobile editing applications like InShot and Adobe's Rush. There are many successful ways to approach the task of editing your social short but the experience of industry practitioners suggests cutting the video and accompanying audio first to create the structure of your story. Remember at this stage the narrative arc for social shorts that takes the audience straight into the action with strong images, statements and quotes.

The duration of videos created for social media varies but we are dealing here specifically with social shorts and there is a consensus amongst producers of around one minute. Be tough with yourself and ensure that your story is lean, packed with strong images, impactful quotes and vital information. With your video cut to length, you can add a suitable music track and mix it with the audio from your contributors.

Captions

The inclusion of captions to provide narration, information (e.g. lower-thirds and credits) and subtitles for interviews or pieces to camera, is a staple feature of social shorts. It is a good idea to create a different style of caption for each of these purposes and every interviewee, so that somebody watching but not listening receives the visual signal that there is a change of voice. This could be as simple as a different colour, bold text, new font, use of a background or a combination of these options.

You could use white text on a dark
background for narrative captions

Contrasting that with black text on
white for subtitling an interviewee

Because we are learning about journalism and not graphic design, we are not expecting elaborate captions and you should focus on keeping them concise, accurate and easy to read. Use the template captions provided by most video editing systems as a starting point.

There are some furthers considerations when it comes to creating captions. The graphics generators in editing systems rarely include a spellchecker, so be extra careful with your spelling and copy your caption text into an application that does spell checking if you need the added assurance. Also be aware that you do not have to transcribe interview clips word-for-word, so long as you faithfully subedit or paraphrase. Do not include fillers such as ‘um’ or ‘erm’.

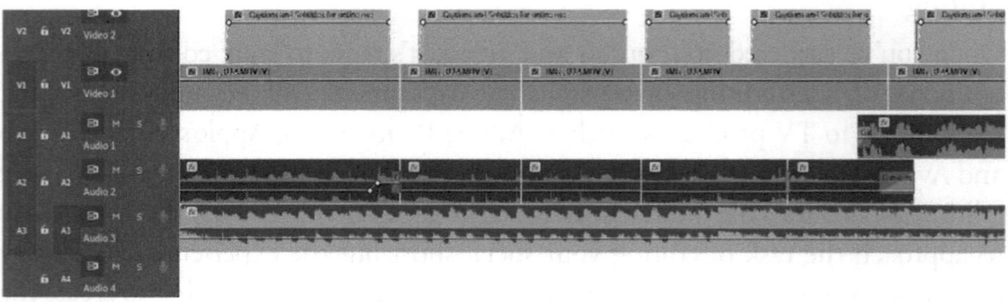

Figure 1.3: A timeline in Adobe Premier Pro for a social short.

The example social short video timeline in *figure 1.3* illustrates how frequently captions appear on screen in these videos (track V2) and how a musical 'bed' (track A3) runs throughout the piece. It's worth pointing out here that captions, as well as their practical purposes, are thought to help retain audience attention. The digital team at Channel 4 that covers sport, was briefed by a Meta representative at a presentation to use captions to keep viewers glued. Jack Gilmartin, who works for this department, says they were also encouraged to 'word flash', i.e. put one word on the screen at a time to keep the text constantly active and provide a further visual lure. However, the style of textual presentation should reflect the tastes of your platform audience and the nature of the story you're telling. Big pop art captions that flash on the screen might work for a music story on Instagram but be a poor fit for a serious news story that's going to Facebook.

Producing your first video of this kind requires a lot of planning and production. The good news is that once you've created your first social short edit, you can preserve elements of it to function as a template for future efforts. Consider creating a second timeline within the project you used to make your social video and copy across the captions, music, end card and any other media that could be reused. Save that timeline to get off to a quick start in post-production the next time you make a social short.

Recap

As you've seen in this subchapter, there is a lot to making a social short video. So, here's a summary of the essentials that you can tick off as you make your film and refer to for revision in future.

Selecting your story

- Choose one that will have an emotional impact through a relatable case study.
- Look for powerful images to help your video stand out in a crowded feed.
- Pick subjects that appeal to younger audiences, such as educational issues, social injustices and environmental news.
- Quirky stories are also popular and fit with platforms that entertain as well as inform.

Telling your story

- Choose one story and do it justice, focusing on the person who brings it to life through sharing their experience.
- Grab the audience's attention from the start with powerful images and quotes.
- Keep up the intensity with more hard-hitting content to retain audience attention.
- Exposition is important but leave it until after the hook of the human-interest angle.

- Invert the traditional TV narrative arc: Action – > Exposition – > Further action – > Conclusion.

Making the video

- Film in the vertical orientation if you're certain your video is solely for digital sharing.
- Use appropriate mics to record clean audio and find a suitable instrumental music track.
- Edit the video first to ensure that your pictures and clips tell the story as far as possible.
- Storyboard your captions using a text application and use formatting options to distinguish between voices.

> **Exercise 1.2: Make your own social short**
>
> There is no better way to test your understanding of social shorts than producing one yourself. As a first attempt, base your video on a day-in-the-life. This will encourage you to focus on the kind of personal story that appeals to a social media audience. Find somebody who does something interesting, unusual, moving, inspiring or thought-provoking and film them doing it using the mobile journalism filming skills taught in Chapter 6. Then use the guidance supplied in this section to turn it into a social short video.

Creating from TV footage

Although social shorts are sometimes edited from specially shot footage, in the case of broadcasters, they are often created from video that was primarily gathered for TV. Editors say that in the early days of digital production, materials would more-or-less be taken from their native platform and posted on a social channel with no adaptations. So, for instance, whatever words might have been written as the cue for a story that appeared in a TV news bulletin, would be the same words used in a Facebook post, that would also include a made-for-TV video of the same story. Now, that same TV content is deconstructed and rebuilt with social media and mobile sensibilities in mind.

At Sky News, this is a specialist job as they employ a team of around seven people whose principal role is to repurpose TV footage for social short videos. A BBC digital editor estimated in 2019 the proportions of sources of video used on its digital platforms: 'I'd say at the minute 70 % TV, 20 % self-shot and 10 % UGC (user-generated content)'. That is likely to have changed as newsrooms increase their understanding of the differing demands of a social media audience.

ITV News has reported that they are looking to do more of their own filming and where possible, ask TV colleagues to pick up shots that specifically meet the needs of digital production. In common with other social short producers who work for broadcasters, they usually cut their video from the rushes (unedited, raw material) rather than the TV package.

But learning how to repurpose video shot for TV is likely to remain a useful endeavour, both because it will remain common practice and it helps develop your storytelling skills as you craft the narrative arc in different scenarios. The task is to reimagine the TV report for a social media audience, keeping in mind the qualities of a successful social short video that we've covered. Jacob from the BBC says that he'll watch the edited TV package, looking for the most impactful material, so that he can place it at the beginning of his social short: 'the most engaging quote, or sequence or shot – straight at the top'.

The approach to editing the piece is the same as for a specially shot video with one key difference: after editing the pictures, that will have been shot horizontally, you'll need to convert them. Vertical video comes in a few shapes and sizes, but generally with an aspect ratio of 9:16, the opposite of TV. Video editing systems will allow you to change the whole sequence to vertical in one go. For example, with Adobe Premier Pro, go to the 'Sequence settings' menu and overwrite the values in the 'Frame Size' field so that the aspect ratio becomes 9:16. For example, if the original video is horizontal 1280 x vertical 720, switch those numbers around so that it's horizontal 720 and vertical 1280.

There's still a bit more work to do because this change of setting gives you a letter box effect with black margins either side of your video. Use the scale, position and crop tools in your editing application to remove any unwanted margins and adjust the image so that the most important part of the picture is clearly visible. Do the same to all the video clips on your timeline and watch back to make sure the framing works all the way through. With the video in the correct orientation, you can now complete the social short by adding captions.

Exercise 1.3: Design a social short from a TV report

Analyse a TV feature or documentary and think about how you could repurpose it for dissemination as social short video. What story would you tell and how does it align with the editorial and production criteria set out in this section? Storyboard your idea by setting out the images, sounds and captions that would appear on the timeline at each stage of the video.

If you've previously produced a TV news report on your course, rework it as a social short, following the guidance given in this chapter.

Dissemination

Where people access news influences how it's disseminated and there is evidence that social media is increasingly the gateway, as reported by the Reuters Institute: 'Across all devices, our data show direct access to apps and websites becoming less important over time and social media becoming more important, partly due to their ubiquity and convenience' (Newman et al., 2022, p. 23). All the major social media sites enable you to share social short videos, giving you lots of opportunities to distribute your work. As a journalism student, you could use your own social channels or those of your institution or media society. Talk to your tutors about the best approach and make sure that somebody has checked your work for any potential legal, regulatory or ethical issues before you post.

The Poynter Institute advises getting to know social platforms and how they promote posts, also suggesting that you familiarise yourself with the users of the platform and try to align with their interests (Bean, 2021). They suggest doing this by 'watching content on the platform you're producing for. By watching, you can pick up current conventions and cues so that your content can feel in touch with the current atmosphere'. There is much more on knowing the platforms in Chapter 5.

As well as sharing on your own account, you could also post your social short to community and fan pages that host people with a particular interest in the topic. Ryan Daniels, a Social Media Producer from CBS Sports, endorses this approach, saying, 'if we've done a video on Manchester United, we can extend the reach by going on to Reddit and posting it on a Man U Reddit forum there or go onto Facebook and include the link on there to our YouTube channel'.

Georgia Carroll, a Digital Producer at Channel 4, says they keep an eye on who and what is trending. This can give you ideas for stories in the first instance but also highlights hashtags that are commonly used for the subject you're covering that you can include in the post that accompanies your video. Jacob from the BBC says that any stories that have a wide impact are good for shareability, with weather events as an example. We all care about happenings that impact on us and weather stories are also likely to affect friends and family in the same region. Another tip from our industry specialists comes from social media marketer Laura Walcott, who advocates dissemination by saturation. She looks to post her content to as many suitable platforms as possible and will repost the same material more than once to increase the chances of it being shown in social feeds.

If you're unsure where to upload your work, enjoy the freedom to experiment and see where you get the best response. Timing is important too. You'll notice an uptick in the amount of content, including social short videos, that publishers and broadcasters share early in the morning, as people get ready for work, and late afternoon, as they travel home.

Summary

To get your social video seen, think about the following: familiarise yourself with the platforms and what makes popular content on them; promote on relevant fan or community pages; use trending hashtags; time you post for when people are checking their phones; post more than once and across all applicable platforms.

Video explainers

What is a video explainer?

USA digital native Vox has led the way in producing in-depth video explainers, lifting the lid on important aspects of big stories from around the world. A standfirst on their website makes their rationale and objectives clear: 'We live in a world of too much information and too little context. Too much noise and too little insight. That's where Vox's explainers come in' (Vox, 2022). In other words, this format is a species of public service journalism, clarifying matters of social, political and cultural significance, so that citizens can make informed choices.

Video explainers that are designed for social and mobile distribution vary stylistically with some including social short-type captions and others that do not. But one consistent feature is that they align with the news habits and attitudes of younger audiences, who are interested in stories that are visual, easy to consume and provide context (Newman et al., 2019, p. 57). Vox produce a lot of graphics-heavy pieces with animations, video, UGC, still images and a voiceover for narration. In common with other producers of video explainers, the tone of their work is unlike traditional news output. Although the topics are often serious, they are addressed informally, with levity and a pacey edit that often includes an upbeat soundtrack. This approach seems to cater for the desire of 18 to 34-year-olds to have news that is both meaningful and entertaining (Newman et al., 2019, p. 58).

UK-based digital editors have picked up on the popularity and purposefulness of explainers. Newspaper group JPI Media sees this as a revival of public information journalism that is 'being brought into the digital age more and more'. ITV News has developed a similar approach to Vox, producing video explainers aimed at younger viewers, though their style is to have a presenter talking to us down the barrel of the lens, leading the audience through the context for a big story, as you can see in *figure 1.4*. They refer to it as evergreen content that holds its value: 'With a Brexit explainer, we can reshare that and repost that most likely for some time to come'.

Figure 1.4: Natalia Jorquera explains climate neutrality in an ITV News explainer.

The Financial Times, a newspaper in origin, creates both presenter and graphic-led explainers and they've found that the format does well on social media, consistently garnering strong viewing figures and engagement. In Australia, Channel 9 News has produced and promoted explainers on issues and processes during the country's own federal elections and as part of their coverage of important overseas ballots. So, if a newspaper that's been around since 1888 and TV companies that have been around for decades are making these things, it's fair to say that they've caught on. Similar to social shorts, this is a format that is internationally popular and here to stay. It is therefore another news format that you should be familiar with and practice creating.

Story selection

The subject matter of explainers often relates to heavyweight political stories such as elections, conflicts and controversial law changes, but contextual pieces are also produced on sports, science and cultural oddities, like why rollercoaster loops are no longer circular (Vega, 2022). In her guide to creating explainers,

Meena Thiruvengadam says that the required competencies are some of the familiar fundamentals of journalism, such as 'savvy news judgement, inquisitive reporting and the skills to tell a strong story' (2013). Though topics vary, the aim is always to plug an information gap of some kind, a barrier that stops people understanding what's happening in the world. When selecting stories and going about the business of preparing your explainer, consider the following:

- Myth buster: What's happening in the world that people have a misconception of or where misinformation may have taken hold?
- Jargon buster: You could directly aim to demystify the language associated with big stories by explaining the key terms.
- Profile: Is there an individual who keeps cropping up in the news whose background you could profile in an explainer?
- Assumed knowledge: A report on interest rates rising will not mean much to viewers who are unclear about what interest rates are and why they matter. If you spot a gap between what journalists assume and the public understands, close it with an explainer.
- Recurring themes: Identify information that would help the public understand a variety of stories on the same subject, e.g. a piece on the break-up of the Soviet Union would work alongside many stories about conflicts and disputes in the region.
- Trust your own instincts: If you're reporting on an issue and are unclear on a key concept, trust that your audience will feel the same. Do the research, learn about it yourself, then make an explainer video.
- Friends and family test: Ask people you know, who are not professionally involved with news and current affairs if, for example, they know what 'fracking' is and why it's controversial? If not, that is a good sign that an explainer on the topic would be of interest to your audience.

Storytelling

Now you've decided on the subject of your explainer, how do you turn it into a story that people might be interested in watching? Explainers have the potential to be dry lists of facts and data. How do we avoid that and instead create an engaging narrative? The BBC's Ros Atkins came to prominence during the coronavirus pandemic in the UK with explainer videos that cut through political spin and specialist scientific language. Instead, he focused on the facts, informing the UK about the progression of the virus and the impact of decisions being taken by government. Many went viral. He says that it is essential to map out the story in advance in a way that a film director might storyboard a movie before shooting begins.

As with social shorts, practitioners recommend starting with strong material by leading with the most important or surprising aspect of the information you've gathered. You might do that by asking a simple question that your audience is likely to want the answer to: 'So, why are fuel prices rising so fast?' After grabbing our attention, take us on the journey towards understanding the issue you've identified. Go through the evidence you've found and highlight any complexities or contradictions along the way. Atkins emphasises the need to make the report flow and he uses 'a technique of scripting where you're constantly hooking off the back of something and throwing onto the next thing' (Adami, 2022). Another essential is reminding viewers why they should care and he often uses the phrase 'this matters because..'., to make key points explicit.

More than ever, avoid jargon or technical language that might obscure the facts from your audience. Say it directly with words that you and they are comfortable with. Finally, come to a resolution that encapsulates what we've discovered. In other words, deploy the abiding fundamentals of narration; sell us the story, deliver on what you've promised and make clear what we've learned at the end.

Charts, diagrams and graphics are used to bring information and statistics to life in some explainers. Have a look at the section on data visualisation in Chapter 6: Journalism skills for the digital age.

Summary

The ingredients of good storytelling for explainers: storyboard in advance; start with strong material; create flow by writing transitions between information; make key points explicit; avoid jargon and technical language; reach a resolution.

Production

Presentation

As we've seen, there's more than one way to go about producing a video explainer. They vary greatly in duration from snappy one-minute pieces for TikTok, to forensic analyses of complex topics that can be around ten minutes in length, such as the films created by NowThis for their Originals series on YouTube. Presenter-led and graphics-based approaches are both popular. As a first attempt, to help you focus on the journalism rather than the technicalities, have a go at a presenter-led explainer, telling the story with the help of video, photos, UGC images and basic textual graphics.

Chroma keying, using a green screen or blue screen, could be used to incorporate relevant images and text behind the presenter, as illustrated in *figure 1.4*, where ITV reporter Natalia Jorquera presents an explainer on the meaning of *net zero*. You can cut between the presenter-led elements to sequences of shots and interview clips. Note that if you choose this approach, it's even more important to map out the video in advance. You need to know which bits are recorded in vision, which will have voiceover and what images, graphics and audio you'll use along the way. There are lots of video tutorials available online that will lead you through the process of superimposing your presenter at the green screen onto images that illustrate the report. Have a look within the online training resources provided by the publisher of your editing software or search in YouTube for 'chroma key' and the name of your editing application.

A twist on this method is to take your presentation out-and-about and film the presenter in relevant locations, recording links a bit like you would shoot pieces to camera for TV. Alternatively, for the camera-shy, you could make an explainer using images and captions in the manner of a social short video. Edit such an explainer using the approach described in the social short subchapter.

Research

As ever, the substance of your journalism is more important than the style of video you choose to produce. Have your audience and the news agenda in mind as you select a topic, process or phenomenon to illuminate. Rigorously research the subject until you're happy that you fully understand what you'll be talking about and then map out the structure for you explainer.

Narration

Remember the advice from Ros Atkins to pay particular attention to how you connect the facts together and return to the question of why it matters. Make sure that when you plan your story you consider the visuals as well as the script, deciding what video, photos, interview clips and graphics you'll use. Do not skimp on the ending. It is something that we often see from inexperienced reporters who load a lot of effort into getting their piece started to the neglect of how it will conclude. You've to know where you're going before you set off.

Storyboarding

The table illustrates one approach to planning your explainer, mapping the images viewers will see alongside the words they'll hear or read. This method essentially creates a paper edit of your video, thereby speeding up the actual editing process when the time comes.

Table 1.1: Storyboard for an explainer on extension of coronavirus vaccine rollout in England, based on an ITV News production

Images	Sounds
	{Instrumental music bed throughout}
PRESENTER IN VISION (in vis) at the green screen with image of vaccine vials behind.	SCRIPT: Under-18? You won't be offered a Covid vaccine unless you're vulnerable.
FULL SCREEN GRAPHIC: Who can get jabbed?	MUSIC
PRESENTER IN VISION	SCRIPT: So it was just adults over 16 and vulnerable to the virus who could get the jab.
SHOTS: Young person receiving jab.	SCRIPT: But now it's been extended to 12 to 15-year-olds who are vulnerable and 17 year-olds who are turning 18 within three months.
FULL SCREEN GRAPHIC: Why has this been decided?	MUSIC
PRESENTER IN VISION	SCRIPT: The scientists responsible for the vaccine rollout have advised the government against offering jabs to all children. They say it's because more safety data is needed and that this age group has low risk of serious Covid symptoms.
INTERVIEW: Professor Chris Whitty, Chief Medical Officer for England.	CLIP: Happy with revised policy and this is the best approach for now.
FULL SCREEN GRAPHIC: Do other countries vaccinate children?	MUSIC
PRESENTER IN VISION	SCRIPT: Yes, countries like the USA and Israel have both begun to vaccinate children.
SHOTS: Children in USA receiving vaccine.	SCRIPT: They've done this to prevent outbreaks in schools that lead to lots of children missing out on their education.
FULL SCREEN GRAPHIC: Will children be included in vaccine programme in future?	MUSIC
PRESENTER IN VISION	SCRIPT: We don't know just yet. The government has not said whether the vaccine programme will extend to children. But the scientists say that the best way to save lives now is to focus on adults.
END CARD with details of social channels.	MUSIC ENDS

Exercise 1.4: Follow the three steps to creating a video explainer

We suggest a three-stage approach to creating a video explainer of your own:

1. Find and study three examples from a range of publishers. Think about story selection, research and production. What common traits can you identify and include in your own work?
2. Choose a subject for your explainer based on the guidance supplied. Research and plan your work then create a storyboard using the simple model in *table 1.1*.
3. Make your explainer. They vary in length but it makes sense to start modest and test that you've grasped the key concepts, so aim for around 90 seconds in duration. Given the work and roles involved, this is an exercise that would suit a team approach. Work together on developing the story but consider sharing out roles such as reporter/presenter, researcher, producer and video editor.

Other types of digital video

TikTok

TikTok burst onto the scene as a serious player in digital journalism in 2020, blazing a trail for super short social videos. It's had a particularly profound impact on younger audiences and 'is the fastest growing social and video network', (Newman et al., 2025, p. 11). The Pew Research Center observed the same rise in popularity, finding in 2025 that in the USA, 55 % of TikTok users consume news on the platform, up from 22 % in 2020 (2025). Research by Vazquez-Herrero et al. claims news organisations began experimenting with TikTok in 2019. The Russia-Ukraine conflict then acted as an accelerant, as Ukrainians used the platform to document their experiences, 'clocking up millions of views in the process' (Newman et al., 2022, p. 25). News producers around the world now invest heavily in making videos for TikTok, hoping to 'add to their presence on social media apps, ensuring that they are reaching all audiences consuming news on the go' (Wu, p. 3).

In line with other social media, journalists have adapted to both the affordances of TikTok and the behaviours of its users, resulting in a news/social hybrid. They have explored new ways of sharing their work, adopting the stylings of the platform whilst upholding the fundamentals of their profession. That means combining an entertaining production with substantive information (2021, Bean). The vast majority (85.3 %) of news videos were found to be specially created for TikTok and virtually all at least correctly modified for the platform (Vazquez-Herrero et al., p. 1728). This shows its unique traits and how its videos differ from the generic social shorts that we've looked at previously. For that reason, we've included this subchapter on TikTok to highlight the key differences; the points where TikTok news video production diverges from social shorts in general.

Story choice

The range of subject matter seen in TikTok videos is not so different to what we discovered when analysing social shorts. Perhaps surprisingly, in her study of 200 TikTok videos produced by a range of well-known Western news organisations, Wu discovered that the news values that drove story selection were not so different to those we are familiar with from legacy media:

> Results of this study indicate that "news" as it is traditionally conceived has not been diluted on TikTok to the extent that many might fear, given its short video format and seeming focus on general entertainment. It seems news organizations have leveraged on the dynamic and creative nature of the platform without sacrificing long-established values of journalism in concerning ways.

A familiar mix of stories has been observed in other studies of TikTok content, with news videos that go viral on the platform varying in subject matter along the hard/soft spectrum, including stories about celebrity, animals and current social issues (Vazquez-Herrero et al., 2022, p. 1724), along with anything quirky and/or humorous. As explained in this section, it's often the way you tell the story that sets TikTok journalism aside from traditional forms, rather than the subject matter.

Storytelling

The extra short duration of TikToks is one of their key characteristics. Videos published by news organisations on the platform are typically 22 seconds long, ranging from six seconds to one minute in duration (Vazquez-Herrero et al., 2022, p. 1724). Georgia from Channel 4, adds that 'the ideal length is closer to 10–20 seconds –with sixty being the maximum when it comes to audience attention span'. Storytelling is pacey and concise. Videos need to capture attention and production is key to achieving this. Georgia suggests 'using moving graphics, large, colourful subtitles and effects. Make sure there is enough going on on the screen to keep the viewer from instantly scrolling on. Start with an outlandish soundbite, so people are hooked to watch on'.

As with social shorts, keeping your audience after you've attracted them in is also essential, as Georgia explains:

> Retention is even more important on TikTok; you want the video to loop, so you want the viewer to stay to the end. Viewers know they can scroll down and most likely find something more entertaining, so you have to make sure you're clipping up the most impactful moments to keep them from scrolling and end the clips on the best moment, so viewers don't get bored or leave too early.

Presentation

A pattern of organisations using the same presenter to front their videos each time, or choosing from a small pool, has also been observed, promoting familiarity and creating a personal bond between the viewer and the producer (Vazquez-Herrero et al., 2022, p. 1723), as befits a platform that is home to influencers and TikTok celebrities. Approaches to how the presenter appears in the videos differ but include having them in the foreground with a virtual background containing relevant images (a bit like a green screen recording), or using the split screen approach, with the presenter in one frame and the images they're talking about in another.

Both styles of presentation are popular because they make use of the available tools; the in-app *Duet* editing function allows users to respond to existing TikToks, creating the split screen effect, and the *green screen* option is labelled as such in the app. In *figures 1.5 and 1.6*, we see the contrast between BBC journalist Alice Cruickshank appearing at the bottom of the screen, Duet-style, with images that help tell her story in the box above. On the right, Guardian reporter Madeleine Finlay has a virtual background created using the green screen function.

Figure 1.5: Journalist Alice Cruickshank appears Duet-style in a BBC News TikTok.

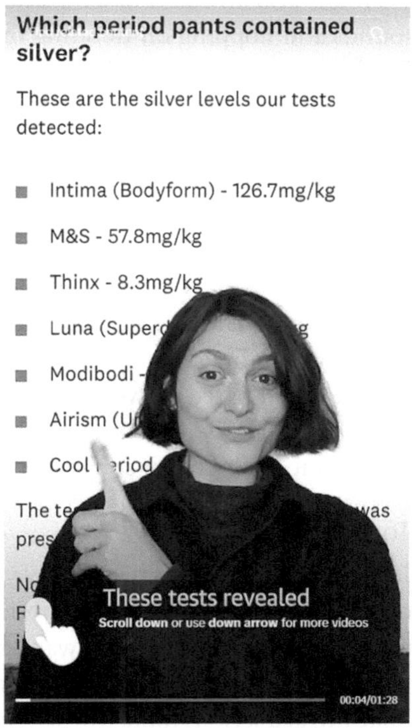

Figure 1.6: Guardian reporter Madeleine Finlay using green screen for her TikTok.

There are lots of guides on YouTube and elsewhere on the internet that explain the simple process of creating a Duet or green screen. However, you could use a conventional video editing software to combine the presenter and content, giving you more control over the edit, such as the additional ability to cut from the shared screen to important clips that fill the frame. Using shared screen presentation is a good example of how the characteristics of a social platform, familiar to and popular with its users, can be harnessed by journalists to tell their stories.

Another useful feature in TIkTok for journalists is '*Stitch*', which is a way of attaching your video to one that already exists. You can select up to five seconds of an existing TikTok, so long as the creator has the function switched on in their settings. This clip then plays before the video that you've produced, allowing you to pick up off the back of it and tell your story. You might, for example, use it to identify and investigate an example of misinformation spreading on the platform, taking a sample section of the kind of falsehoods being shared and following it up with your evidenced and well-sourced investigation to bust the myth.

Audio

TikTok is also an audio-based platform, where videos go viral for using specific sounds and hopping on trending music. Unlike other social platforms, such as Facebook, viewers almost always use the app with audio playing. Georgia suggests, 'you might want to lead with the best sound bites, in the hopes that the sound might be repurposed by other users'. Making these connections through sound is integral to TikTok and news publishers have shown a varying degree of engagement with that characteristic. In one study, most news videos (61.4 %) missed the point by including original sound and that had consequences for views: 'Since it was not integrated from the TikTok music library, it was not linked to other publications that used the same song and limited the circulation of the video through sound tagging' (Vazquez-Herrero et al., 2022, p. 1726). Those who did use the sound-based networking affordances, the Washington Post chief amongst them, had success responding to trends and viral challenges and by using the TikTok music library or songs that are popular on the platform.

Tone of voice

It's not just what you say, it's how you say it. The tone of TikTok is different to other social media. Whereas emotional, human stories fit well with Facebook, the younger TikTok audience wants to be entertained so, as Georgia explains, it's 'much more humour based'. That can be challenging for journalists. How do you poke fun at a serious topic whilst being informative and without causing offence? The Washington Post has had a lot of success in this area, hiring charismatic

TikTok producer/hosts who cover heavyweight stories through performance, with the satirical edge you'd associate with comedy sketch shows like Saturday Night Live. One of the hosts, Carmella Boykin, regularly produces skits in which she plays the roles (all of them) of senior officials (Federal judges, senators, CEOs, etc.) arguing over the political and legal controversies of the day. Humour is a matter of taste, but we find the videos amusing. Crucially, they're also informative, keeping up to date with breaking news and reflecting the different positions taken in a dispute, albeit with levity.

Reach

Producing videos for TikTok is one thing but getting eyes on them is another. The recommendation algorithm has more influence over what users see than who they follow. Making sound connections is one way to increase reach, as we've discussed. Vazquez-Herrero also suggests hashtags, lip syncs and creating challenges (p. 1729). The importance of hashtags is emphasised, with nearly half of the videos in their sample including trending hashtags (p. 1724). It's worth pointing out that although reach is clearly important, because we want an audience for our journalism, news organisations also see value in simply being visible to younger audiences.

Shorts and Reels

Some social media channels have developed video formats that are, essentially, social shorts. For example, YouTube Shorts was launched in 2021 in response to the growing popularity of TikTok. This section of the YouTube site caters for vertical videos, for compatibility with mobile consumption, that are up to one minute in duration. Creators can produce them in-app using licenced music, text, graphics and basic editing options.

Instagram 'Reels' emerged in 2020 and were similarly 'inspired' by TikTok. It signalled a pivot from the platform towards video content and now, like YouTube Shorts, it's difficult to find a legacy or digital-native news organisation that does not routinely publish Reels on their Insta accounts. Even the New York Times, that slammed the new video form when it was launched (Chen and Taylor, 2020), routinely gets millions of views for its Reels.

In the UK, The Guardian's Reels also do good numbers. Little, if any, of the video they use is specially shot, as they instead draw from UGC, agency pictures and other forms of syndicated footage, to use in conjunction with their own graphics. Spain's El Mundo takes a similar approach, which is understandably common amongst newspaper producers, who want a piece of the visual journalism action but don't have a legacy fleet of camera crews and video journalists.

The maximum duration of a Reel recording defaults to 15 seconds, though you can adjust this setting up to 90 seconds. The rules seem to be getting more relaxed over time, seemingly heading to the point where you can pretty much publish any sort of video, so long as it's in the vertical orientation. Reels can be made in-app using similar tools to YouTube Shorts. You can add audio from the Instagram library or choose your own music and apply video filters and effects, such as speeding up or slowing down your shots.

With both Insta Reels and YouTube Shorts, you can also upload videos that you've already created, before using the available in-app post-production features. This is an important efficiency for digital journalists. If, using an editing system that you're familiar with, you can produce a short that is a good fit for more than one social network, the process of then making small adaptations post-upload is much more efficient than producing lots of bespoke, in-app versions of the same story.

References and recommended reading

Adami, M. 2022. *The art of the viral news explainer.* [Online]. [Accessed 5 July 2022]. Available from: <https://reutersinstitute.politics.ox.ac.uk/calendar/art-viral-news-explainer>

Bean, A. 2021. *Crafting a video content strategy for social platforms.* [Online]. [Accessed 16 June 2022]. Available from: <https://www.poynter.org/reporting-editing/2021/video-content-strategy-social-platforms/>

Harcup, T. 2015. *Journalism: principles and practice (3rd edition.).* London: SAGE.

Negredo, S., Martínez-Costa, M-P., Breiner, J., and Salaverría, R. 2020. Journalism Expands in Spite of the Crisis: Digital-Native News Media in Spain. *Media and Communication.* **8**(2), pp. 73–85.

Newman, N., Fletcher, R., Kalogeropoulos, A. and Nielsen, R. K. 2019. *Reuters Institute Digital News Report 2019.* Oxford: Reuters Institute for the Study of Journalism.

Newman, N., Fletcher, R., Robertson, C-T., Eddy, K. and Nielsen, R. K. 2022. *Reuters Institute Digital News Report 2022.* Oxford: Reuters Institute for the Study of Journalism.

Newman, N., Fletcher, R., Robertson, C-T, Arguedas, A. R. and Nielsen, R. K. 2024. *Reuters Institute Digital News Report 2024.* Oxford: Reuters Institute for the Study of Journalism.

Newman, N. R, Fletcher, R., Robertson, C-T, Arguedas, A.R. and Nielsen, R. K. 2025. Reuters Institute Digital News Report 2025. Oxford: Reuters Institute for the Study of Journalism.

Pew Research Center. 2025. *Social Media and News Fact Sheet.* [Online]. [Accessed 12 January 2026]. Available from: <https://www.pewresearch.org/journalism/fact-sheet/social-media-and-news->

Thiruvengadam, M. 2013. *How journalists can create better explainers.* [Online]. [Accessed 5 July 2022]. Available from: <https://www.poynter.org/newsletters/2013/how-journalists-can-create-better-explainers/>

Vázquez-Herrero, J., Negreira-Rey, M-C. and López-García, X. 2022. Mind the Gap! Journalism on Social Media and News Consumption Among Young Audiences. *International Journal of Communication.* **16**, pp. 3822–3842.

Vega, E. 2022. Why roller coaster loops aren't circular anymore. *Vox.* [Online]. [Accessed 6 July 2022]. Available from: <https://www.vox.com/videos/2022/6/29/23188504/roller-coaster-loops-circular-g-force>

Vox. 2022. [Online]. [Accessed 5 July 2022]. Available from: <https://www.vox.com/explainers>

Walker, M. 2021. *U.S. newsroom employment has fallen 26 % since 2008.* [Online]. [Accessed 8 July 2022]. Available from: <https://www.pewresearch.org/fact-tank/2021/07/13/u-s-newsroom-employment-has-fallen-26-since-2008/>

Wu, S. 2025. Is TikTok News Less "News-Like"? Investigating News Organizations' Stories on TikTok and Their Alignment/ Divergence with Traditional News Values, Norms, Routines, and Roles, *Digital Journalism.* pp. 1–19.

Social storytelling

The way that people access news worldwide has changed, with audiences for print and broadcast journalism declining as younger groups prefer 'to access news via side-door routes such as social media' (Newman et al., 2022, p. 11). Journalism has adapted accordingly, increasing the volume of reporting available on social networks. In the early days of this evolution, news organisations and reporters tended to use these third-party platforms to promote their own digital offerings, with links to their websites where they control the content, data and revenue streams. Indeed, this is still how it is often done.

But the news media has also taken to creating native content. By that, we mean that they are publishing posts of different kinds that report the story within the platform, using the available tools and media affordances, rather than linking to their proprietary websites. The social short videos we focused on in Chapter 1, are an example of this native content. Ryan, from CBS Sports, argues that it suits audience behaviour: 'when you're on X, you're usually on X. You don't tend to click off X to go on to another site, and if you're on YouTube, you're only on YouTube'.

Jonny McGuigan, Executive News Editor for Growth and Social at the BBC, has oversight of many areas of the corporation's digital factual output. He says they have mixed feelings about creating native content because of a lack of control over third-party platforms, limited access to the associated data and uncertainty over the income it will generate. Because of this, strategies around native content creation have swung one way then the other. But it is clear, not least with the arrival of TikTok, that news organisations now invest heavily in creating stories and sharing them within social media, with no referral to their own platforms. The process of creating this content is what we will refer to as 'social storytelling'.

Putting the historical and economic issues to one side, social storytelling is about creating and sharing journalism using the many ways of conveying information provided by social media. We focused on digital videos in the first chapter as they are such a popular and prevalent form of native content and have been for a long time, around the world and across the full range of journalism producers. In this chapter, we'll show you the best of the rest. You'll discover a range of storytelling formats, receive production guidance, learn how to match stories to formats and have a go at creating your own social stories.

Exercise 2.1: Familiarise yourselves with social storytelling, starting at home

To get you thinking about what stories you might tell and how you'll tell them on social media, particularly to the social-native audience you're a part of, scroll through your own feeds for five minutes and make a note of the content that stands out to you. What made you stop and take notice? Was it the story, the way it was presented, or a bit of both? Not everything you find will be journalism but it will give you a clue as to what stories and story formats work for a social audience. Can you identify storytelling, presentation, filming, editing and audio techniques from the work you see that you can bring into your social media journalism?

Formats

Numerous story formats have emerged since journalists took up the challenge of creating native social media content as a way of distributing stories. The list is long and regularly being added to as reporters adapt to new platforms, production tools and presentation trends. It is unnecessary, not to mention impossible, to cover them all here, so we'll focus on examples of some of the most common styles that have become a staple of digital news production. The emphasis is on generic formats and production principles, though we will illustrate at times by referring to specific functions within existing social media. It's inevitable that these platforms and their tools will change over time, but it is also likely that the key traits of the format will persist. For example, there will always be ways of telling stories using photos and words in combination. The formats we have chosen are commonly available across platforms and have been used by social media journalists around the world for many years.

Image and text

One of the simplest formats is a photograph accompanied by a short post. Between them, they supply readers with the top line (the crucial first sentence that explains what the story is about) and a picture that shows what's happening. A striking image and a strong top line can have a lot of impact and substance, drawing the eye of the social viewer and telling them the story of the event in a concise, snackable fashion.

In the example *figure 2.1*, a sports news story is broken on X (formerly Twitter) by freelance journalist Fabrizio Romano. The quote from England football manager Gareth Southgate in support of his player Harry Maguire is reflected in the photo. Maguire's selection was contentious at the time, but that controversy is left unsaid in this specific post. Football has a huge following in the UK, so it's not surprising that this post got such a strong reaction from readers, with 2,648 comments and more than forty thousand additional engagements through reposts and likes. This shows that a short and simple story, told natively through an image and a couple of sentences, can go a long way.

Figure 2.1: Fabrizio Romano breaks a football story on X.

Production

Uploading an image from your mobile phone and adding a few words is a straightforward production task, so the key here is to spot the opportunity to create such a post. Even if it's not your first intention to produce a post of this kind, always think photographically when you're out gathering a story for radio, TV or digital video. Look out for scenes, people or objects that reflect what is happening and take a snap to share on social media, accompanied by words that explain the story succinctly. It is an efficient way to reach additional audiences instantaneously and get them interested in your journalism.

An important tool available on many platforms is 'alt text', short for alternative text. It allows you to describe the image and add metadata, such as copyright information, without obscuring the picture. You insert the text at the point of upload and it appears behind a discrete badge that the user can click on when they see your post. It's essential for accessibility as the text you insert is read aloud by screen readers used by the visually impaired. It also contributes to search engine optimisation (SEO), the process of making it more likely people will find your page or post, as search engines will index your additional information about the image.

Video clip and text

The same smartphones that allow reporters to take photos relevant to the story they are telling, can also be used to capture video clips to achieve the same aim. A single clip can be recorded in seconds and instantly shared on social media as a way of telling or updating a story, with no recourse to linking to a title's website. As with photo-based posts, text accompanies the images to explain to viewers what they are looking at and why it matters.

In *figure 2.2*, a video clip of a woman in Northern Ireland making her feelings clear about a murder, was taken by a reporter who's usually in front of camera. It's promptly shared with a brief post and receives more than five thousand engagements (likes and reposts), including 105 comments. The figure also shows a reporter from a USA broadcaster speaking to camera (selfie-style) in Moscow, as Russian security officials clamp down on demonstrations against the invasion of Ukraine. He points out and explains who is in the city square, what they are doing and how it reflects ongoing events. Both these video clips, taken at the scene and made immediately available, throw social media users into the heart of the action.

Production

Unlike filming sequences of images for a video report, with a single shot for social you can keep the camera rolling, capturing the action, until the action is over. This

Figure 2.2: Reporters from ITV News in the UK and ABC News in the USA sharing video on X. The clips have been viewed nearly a million times between them.

means that you have one video clip that can be immediately disseminated on your social channels without the need to edit a string of shots together. Compose a shot that captures the important action and let the story unfold within the frame. Glance away from the smartphone screen to see if there's anything else happening in the environment that you should film for the story and, if you identify anything, move the lens in one confident motion to this new scene, keeping the recording going.

As with the use of a single image, the text should be concise but engaging and informative, providing any contextual information the viewer needs to make sense of the video. In *figure 2.2*, Stephen Murphy tells us what is happening, where, why and who's involved in a single sentence: 'Emotions running high in Derry as friends of #LyraMcKee protest with red paint at HQ of Saoradh, linked with the New IRA'. James Longman uses thirteen short words in *figure 2.2*, with further important details spoken in his report to camera.

Multimedia and text

Social channels often allow users to upload more than one image, or piece of media, to a single post. This leads to the creation of a sequence that helps to tell the story somewhat in the style of a storyboard or slideshow. Instagram creates a carousel when you upload multiple pictures and videos to a single post, with the first image acting as the thumbnail in feeds. If viewers click on that first pic, they can swipe through the rest to follow your report. This has significant storytelling potential, particularly when you consider the prospect of mixing video, graphics and text slides with the still images.

Two slides from a carousel made up of photos and text can be seen in *figures 2.3*. It is from a Vice UK report into what British people think about the impact of Brexit, eight years after the referendum vote to leave the European

Union. The reporter had powerful images to work with and stark statistics. An Instagram carousel gave them the capacity to make the most of both on a platform popular with young adults who largely voted against leaving and would be interested in its consequences.

Figure 2.3: Slides from an Instagram carousel report by Vice UK on attitudes to Brexit.

Linking together multiple posts to create stories with depth and detail is another way of producing multimedia content with text. These *threads* (*see figures 2.4 and 2.5*), as they are referred to, are a feature popularised on Twitter, now X. Meta launched a new app called Threads as an alternative to X in July 2023, with the brand name clearly flagging the ability to tie posts together to form a narrative.

Production

Your task with a carousel, thread or any other mode of multimedia production is to gather and create a media sequence that tells a story. Take the pictures, video and audio you need when you go out to investigate, then add in any supporting rights-free images, music, UGC and graphics, as you compile your report. Text based images, such as *figure 2.5*, can be produced using any suitable graphics tool, with Canva a good choice because it's free to download, can be used on a desktop or mobile, is intuitive and designed for social media production.

As with any other journalistic endeavour, it will pay to plan your work by researching the story and talking to contributors in advance about what they do and what you'll see in the location where you'll meet them. From there, you can produce a storyboard to direct your media gathering. The storyboard

Figure 2.4: The first post in a thread on X detailing the tactics of a football team.

Figure 2.5: The first three posts in a thread on X about snow leopard population.

template provided in the 'Explainer' section in Chapter 1 would be helpful here too. Inevitably, this storyboard will evolve as you go and that is a positive sign, showing that you can adapt to the reality of the situation.

Once you have all your media, the next task is to put it in sequence for publication using a narrative structure that is similar to a social short. Consider the following three components of the story arc:

1. *Beginning.* As with all forms of social storytelling, in this competitive market it is vital to get off to a strong start. Make sure that the first image in the carousel or gallery is eye-catching. If this opening picture is particularly powerful, you may be able to publish it as it is. Usually, it'll pay to add a few words that serve as a headline, explaining, complementing and adding to the appeal of the image.
2. *Middle.* This is where you explore the story and share what you've found in your investigation. As illustrated in *figure 2.3*, a mixture of images and text works well. A text-only graphic may be needed if you have a lot to say and your words would obscure a picture, or if you want to highlight an important quote, statistic or fact. Additional media, such as video, audio clips and data visualisations can be included on merit. In every case, choose the best medium for what you want to communicate, e.g. a video clip from an emotional case study will be more impactful than their words in a textual quote.
3. *End.* Every story needs an ending. Think about summarising what's gone before, reaching a conclusion and pointing to the future. Finish the story

strong visually as well. Do you have an image that captures the situation as a whole? If you've been exploring a divisive issue, do you have a photo that reflects both sides, such as competing politicians side-by-side on stage?

The specific process for creating a *thread* will differ depending on the platform but the essential steps are as follows:

- Create your first post in the usual way then, instead of hitting the 'post' button, choose to add a second post. This option could be signified with a '+' sign or a textual prompt to 'Add to thread'.
- Make your next post, adding media as you see fit, and repeat the process.
- Once all your posts are in place and you're happy that your report is complete, hit the 'post' option.

It is good practice to number each post to indicate to readers how long your thread is and help them keep track of where they are up to in the story. Make it clear on the first post that this is a thread by including the term and/or, as is commonly the case, by using the thread emoji, as illustrated in *figures 2.4 and 2.5*. Then number each post according to its position in the sequence and indicate the total number of entries in your thread, e.g. for the third post of twelve, put '(3/12)' at the end of your text.

Stories

There are stories and then there are Stories. Snapchat added the Story function to its application in 2013. It allows users to put together a series of images that, supposedly, tell the story of their previous 24 hours. It is a rolling process, with your Story continually being updated with new photos, videos and other media, as those older than 24 hours disappear. Facebook and Instagram, now part of the Meta group, followed suit in 2016, even using the same term Story for their 24-hour, ephemeral content. This expansion suggests the Story function is here to stay and worthy of its place in our list of social formats.

Whether what is produced is genuinely a story (i.e. a narrative in the traditional sense) or more a hotchpotch of random pictures, is open to debate. What we can say is that for journalists who are in the business of telling stories using sequences of media,this format can be useful. The fundamentals are the same as for the other multimeda and text formats we've covered in this section, but the difference is that the Story function is often used for promoting your journalism as well as being a way of creating a report.

News organisations that have taken to including these Stories as part of their digital offering include French digital startup Brut. They often create Instagram

Stories that are saved to their highlights. The subject matter varies greatly and includes sports reports, news roundups, features on cultural events, celebrity gossip, inspiring stories and hard news. What they have in coming is a lively mix of images, sounds, text and interactivity, illustrating the flexibility of this format as a means of creating and disseminating journalism. You can see an example in *figure 2.6*, with a slide from a ten-part story on Mardi Gras celebrations, that asks the viewer a true or false question about the history of the carnival.

Stories aren't recommended to users of the platform via their general feeds, e.g. they don't appear in the 'Explore' section of Instagram. Therefore, they don't attract new eyes to your work in the same way that Reels and standard posts might. However, they're useful for keeping your current followers connected to your account because of the prominent position Stories have on homepages. You can also add hyperlinks within your Stories to specific pages of your website, potentially to an online report on the same subject. Furthermore, if you upload a Reel to your Story that is more than 15 seconds long, it will automatically prompt viewers to click on the screen to watch the full Reel. As illustrated in *figure 2.7*, ITV News has used its Instagram Story to provide a roundup of the day's events, branded 'The Rundown'. Other outlets are doing similar news summaries, and in all these examples, we can see how Stories can be used to promote or condense the journalism you have created across your digital channels.

Production

As with other forms of journalistic native content, forethought and planning is the key to success. Produce a draft storyboard for your report before you go out to gather media. This is vital if, for example, you plan to do a piece to camera (PTC). You'll need to know before you talk to camera roughly where the PTC will sit in your story so that you can pick up from the preceding script and lead into what your viewers will find out next.

Table 2.1 exemplifies how you might set out a storyboard in advance of filming. It's written from the point of view of a journalist who's going out to report on a protest in Washington against a change in USA law that removes abortion rights. Ahead of attending the event, there are some things you know (such as the background to the story), some things you can anticipate (protestors to interview) and other things that you cannot. Base your plan on what you know and what you can anticipate, then update the storyboard before you pull the report together. For example, if the weather became rainy, you might do an additional PTC that references conditions and comment on the impact it's had on the demonstration.

When creating a Story, the production is quicker and slicker if you work with your phone as your camera, so that the images and sounds are readily available

Figure 2.6: An interactive screen from a Brut Instagram Story.

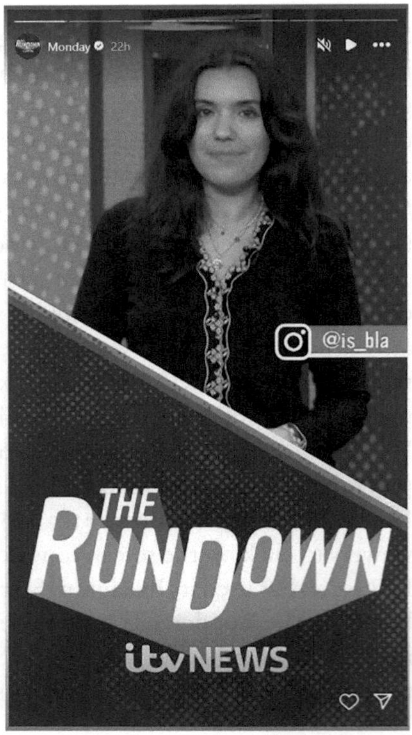

Figure 2.7: A title video to an ITV News Instagram Story series called 'The Rundown'.

in your camera roll when you want to edit and upload them to your Story. You can learn how to use your smartphone for capturing images and sound in the 'mobile filming' section of Chapter 6.

After you've recorded the story, you can modify images using filters, text, backgrounds and stickers, using the tools available in the social media apps themselves or your phone's camera app. You can also create any additional images you may need. In the story depicted in *table 2.1*, you would create an end graphic using the social media app or a user-friendly application such as Canva. Hyperlinks can be created in different ways on different platforms, though it's common to be able to do so by adding a 'sticker', pasting in the link and writing text that prompts the user to click. Tools can also be used to create opinion polls and quizzes if you wish to add those forms of interactivity to your Story, as seen in *figure 2.6*.

With all your media in place and available in your camera roll, the final step is to add them to your Story in the order outlined in your updated storyboard. Again, the precise process will vary between platforms, but the general workflow

Table 2.1: Storyboard for a report on protests against changes to 'Roe versus Wade' legislation in the USA designed for Snapchat, Facebook or Instagram Stories

Images	Words
SLIDE 1: PTC video alongside crowds outside the Capitol	PTC SCRIPT: Behind me are thousands of people, mainly women, who've gathered to have their voices heard. They've come to the Capitol in Washington to pressure lawmakers into thinking again about changing abortion rights in the USA.
SLIDE 2: Wide photo showing scale of protest	TEXT: It's two weeks since the Supreme Court repealed Roe v Wade – the ruling that had guaranteed abortion access for nearly 50 years.
SLIDE 3: Video of protestors chanting	TEXT: Organisers say that more than 10,000 women have travelled from across the country to unite against the court's decision.
SLIDE 4: Video protestor interview	CLIP: Protestor explains where she's from and why she's come.
SLIDE 5: Video second protestor interview	CLIP: Protestor explains where she's from and what she wants to happen now.
SLIDE 6: Close photo of protestor with placard	TEXT: Protestors want President Joe Biden to do more to protect their abortion rights.
SLIDE 7: Photo of Biden	TEXT: Biden has signed an executive order to protect abortion care, but he cannot overrule the Supreme Court, with a 6–3 conservative majority.
SLIDE 8: Graphic and sticker with hyperlink	TEXT: What happens next? We've spoken to legal experts and the people in power. Click here for more on our website.

is to choose the Story option, then hitting 'select' will open your camera roll and you can pick out the images in your desired order.

If you want to save your Story so that it is available for more than 24-hours, on Instagram and Facebook you can permanently pin it to your profile by adding it to your 'highlights'. Snapchat has a similar function; you can save a Story to your 'memories' and reshare it to your public profile whenever you want to.

Social cards

The Financial Times, a London-based newspaper first published in the late nineteenth century, has gained a reputation for being sure-footed and innovative around digital journalism. Their strategy is to match the traditional strengths of their publication with the affordances of digital media. This has led, amongst other things, to the development of what they call a 'social card'. This format imprints the FT's specialism for creating infographics onto visually driven social media, particularly Instagram. An historical economics periodical would not, on the surface of it, seem a good match for a platform that is most popular with 18–29-year-olds, but it works. A digital editor at the FT explains the purpose of social cards:

> To tell a story or deliver a piece of information within that card. So, it's often a graphic or a chart. If it's a graphic, it will be some striking image with a bit of text, and if it's a chart, it'll have essentially a self-contained story within that chart. (Bucknell, 2020, p. 12)

The story of a surge in the value of Bitcoin is visually illustrated with a line graph in *figure 2.8*. The post that accompanies the image adds important contextual information that describes the reason for the sudden increase: 'investors celebrated the election victory of Donald Trump, who has pledged to make the US "the bitcoin superpower of the world"' (Financial Times, 2024). Other publishers and individual journalists have followed the FT's ethos and images that visualise and storify data are common in our social media feeds.

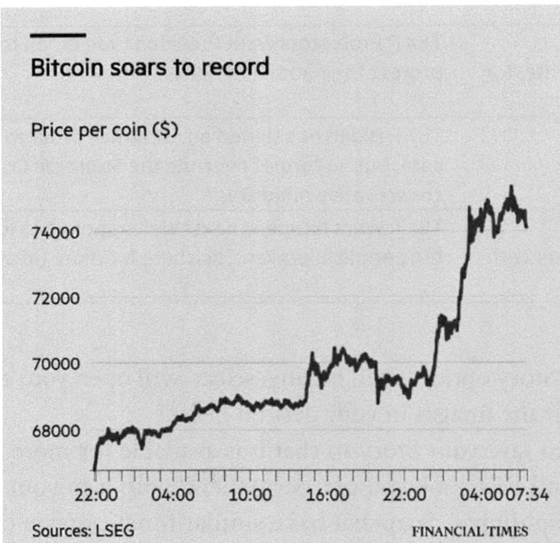

Figure 2.8: A Financial Times 'social card' on Instagram.

Production

This type of storytelling is called 'data visualisation'. We cover the subject in Chapter 6, where we suggest a range of applications, most of them free to use, that are intuitive and make the process of creating infographics more straight forward than you might think.

It's important, as illustrated in *figure 2.8*, that you acknowledge the source of your data in the graphic. This gives due credit to the researchers and lends credibility to your work if it's a known and trusted source. If you have gathered the data yourself, say so in the credit for transparency and so that your readers know that you haven't simply forgotten to state where the information is from.

Funny stuff

We are reporters not comedians but that does not mean we can't have fun and evidence shows that young audiences like a bit of entertainment to go with their news (Newman et al., 2022, p. 43). Therefore, it's not so surprising that memes can and are, albeit rarely, used by journalists to tell stories. Georgia, a digital producer at Channel 4, highlights their shareability as a key motive for considering using them, as they are 'really good for getting people tagging their mates in the comments. It's like, "oh, this is you". 'The Nieman Lab has reported on the journalistic plausibility of memes on the basis that 'they are now key to the communications landscape' (Mina, 2017).

A local TV news station in Denmark has adopted memes as part of their experimentation into which social formats work for young audiences. For TV2 Østjyll, the simple aim is 'to convey the story in just one picture while being funny and easy-to-understand' (Green, 2019). An example of their approach is reporting on an increase in parking spaces with a meme of a car improbably turned on its side between two properly parked vehicles. TV2 Østjyll acknowledges there are challenges; using memes was met with resistance in the newsroom and raised concerns over copyright issues. The station also recognises that they won't be a good fit for every story, not least tragic events, but have found 'topics which people have strong opinions on work the best, as well as those aimed at a wide audience' (Green 2019).

GIFs and emojis can also be found in digital journalism. Have the confidence to communicate in a way that you know makes sense on the platforms you're using and with the audience you target. They may help you to communicate more efficiently (e.g., use a nation's flag emoji rather than its name), make a humorous comment on a story or connect with your audience in ways they understand and expect.

Format choice

As we advocate throughout this book, the production methods are a means to an end and that end is to tell a worthwhile story. Therefore, the story comes first; how you choose to tell it, where you publish it and what format you use follows. That's what we'll consider in this section on choosing the right format for your story, with the following four steps.

1. Find a story.
2. Consider its merits, such as who it will appeal to and how it can be told.

3. Match the story to a platform that aligns with those merits.
4. Gather the material that you need to tell the story.

How that process works is illustrated with examples in *table 2.2* and you can read more about the characteristics and audience demographics of social media in Chapter 5.

Table 2.2: Process of developing a story into native content for social media

Story	Merits	Suitable platform and format	Content
Health problems for students living in houses with mould	Strong visuals of bad conditions, case studies who can testify to problems and large audience of students and parents	An Instagram carousel to make most of images on a platform popular with teens and 18–29-year-olds	Student case study interviews for quotes and/or video clips and photos of mould
First ever 'Light Night' festival in city centre	Striking images and excited spectators, everything you need in one place	A lively Story would work and the parental age profile of Facebook matches this family event	Photos and videos of light displays and crowds, interviews with spectators, PTCs that guide viewers around event
TikTok of disabled man attacked by bouncers goes viral at end of court case	Injustice captured on camera, lots to research	Use 'stitch' function in TikTok to show clip from original video followed by your video, use 'green screen' to report in vision	Interview and shots with victim, right to reply for venue and/or security firm, data on similar cases
USA presidential election is taking place this week	Important global story but the 'electoral college' voting system is complex and needs explaining	An 'explainer video' would work as a reel on Instagram or short on YouTube, reaching a young audience that's politically engaged and wants to understand	Voting system explained by reporter in vision with graphics, expert interview
A protest has developed into a confrontation between rival groups and police	Breaking news story that people will want to know about straight away	A single post on X with a video clip of clashes would immediately announce the news and convey the scale and nature of the trouble to a news-engaged audience	Collect video clips and still images as you witness from safe distance, consider linking together into a thread

There will usually be more than one format and platform that could be a good fit for your story, so the examples in *table 2.2* should be seen as one answer, rather than 'the' answer'. So long as your approach makes a logical match, you're fine. In any case, you'll likely want to share your report across several social channels if you can, so finding multiple matches is a good thing. You'll also notice that we've included digital video from Chapter 1 alongside the formats discussed in this chapter, because they're all examples of native journalistic content.

Story selection

The importance of evoking the feelings of viewers is not restricted to social video. Social content producers regularly cite emotional triggers as being a key ingredient of news going viral no matter the format, having consistently seen stories with a strong emotional pull performing well. Jacob Tomlinson at BBC Yorkshire says he's drawn to human stories, that have a higher probability of eliciting an emotional response. He explains that he tries to get into the mind of the audience by drawing on his own experience, asking himself what would interest him after a day's work when he's relaxing. Story categories that were found to typically elicit these feelings included soft news, such as celebrity and lifestyle, but also uplifting elements to hard news stories, for example an improbable rescue from a collapsed building following an earthquake.

News producers who took part in Vázquez-Herrero et al'.s (2022) European study of social media news consumption amongst 18–25-year-olds, talked about factors that influence story choice for that audience. These include trending topics and subjects that impact on the young, such as education, mental health and job insecurity. A common strategy is to hire young people who are familiar with both the appetites of their peers and the grammar of the social networks. The news producers recognise the challenge of producing work that is both informative and enjoyable in social spaces where they are competing with friends, family, hobbies, interests and entertainment. Besides these topics, the Poynter Institute advocates flipping the lens at times and using social media to take your audience behind the scenes of your reporting (Bean, 2021), making you and your work the story.

Story structure

We've offered specific advice on structure alongside the formats described in this chapter but there are general rules that also apply. As with social videos, the need to grab your viewers' attention straight away is important in all forms of social storytelling. The single-image formats we looked at (social cards and posts with one photo) should only be deployed if your media has that immediate visual impact. In the case of the multimedia formats (e.g. Stories, threads and carousels), the first image and text are vital to hooking the audience. But before they get to that first screen, the thumbnail that appears in search results and account homepages is your first lure. Both the image and text must work together to persuade social media users to click through. You can find out more about how to word your social text in the writing section of Chapter 6.

Although first impressions are key in social storytelling, you do not want to finish on a whimper. A positive last impression of your work, or that of your publisher, will help you gain the trust of your audience and, hopefully, persuade them to subscribe to or follow your account. You could even create an end title that invites them to do just that. Put thought and planning into the ending of your report at the storyboarding stage we have recommended. Summarise what you have shown and said before reaching a conclusion that looks forward.

Exercise 2.2: Follow the steps to create your own native journalism

Demonstrate your understanding of social storytelling by creating and publishing a report natively on social media. As you have seen in this chapter, you have lots of options in terms of the platform, how you craft your post, what story you tell and how you promote your work. Follow these criteria that apply no matter the social media you choose:

- Make sure your story choice, storytelling and chosen platform are aligned (see *table 2.2*).
- Find a story that will immediately capture your audience's attention in the crowded social media marketplace.
- Get your facts right using well-sourced information and provide insight from experts, witnesses, case studies, etc. as appropriate.
- Tell the story clearly using direct and conversational language, and by linking together each element of the narrative before bringing it to a conclusion.
- Use a tone of voice and presentation style that fits both the story and social media site you're using.
- Choose and/or create media appropriate to the platform that enhances the storytelling and makes your work more visually appealing.
- Display your production skills in gathering, editing, processing and laying out the words and media in your story.
- Bring creativity and a willingness to experiment to the task. You know the fundamentals, so do not be afraid to be original.

Dissemination

Many of these social storytelling techniques allow for quick and easy publication, directly from your mobile phone. So it's perhaps worth a reminder at this point that the need for your journalism to be as correct and accurate as possible is just as true in these formats as in any other. Any mistakes in production, spelling or content will reflect on how your journalism is perceived by the audience. It's so simple to press a button and publish your story for the whole world to see but make sure you are checking thoroughly before you go ahead. It's good practice to make sure someone else always looks over your work first as they may spot problems that you haven't noticed yourself.

Many of the dissemination tips offered in the social shorts section (Chapter 1) apply to native content more broadly, but Denisova suggests a couple more approaches to helping your social stories find the greatest possible audience: 'creating dedicated thematic groups [and] tagging influential people and groups' (2022, p. 10). The first recommendation refers to the common practice of subdividing a news organisation's content between a range of specialist pages and accounts, such as a dedicated page on Facebook for science stories, a Twitter account for your business pieces or an Instagram profile focused on fashion.

As an example, Ryan from CBS Sports' social team, says they created a brand called Golazo for domestic and international football content for the USA market. They publish across the full range of platforms under this name, maximising the appeal of their soccer content by separating it from their full range of sports coverage and pushing it to the many fans of the sport.

The second suggestion from Denisova is something that you could try with your stories and will work well if you've been able to secure contributions from people or organisations with large social media followings. At Channel 4's digital sports division, they consider the social appeal of interviewees and groups at the outset. When they cover the England football team, for instance, they target players with a large following amongst young people. Jack Gilmartin, a Digital Producer from Channel 4, told us that an Instagram Reel they created featuring popular winger Bukayo Saka achieved around one million views, a hundred times greater than the 10,000 their videos generally attract. He puts this down to both the appeal of the player and buy in from the England football setup, that also has a massive following.

The words you use to introduce your story in the post, thumbnail or cover, will also impact on how many people look at your work. This opens the delicate topic of the distinction between the low-brow tactic of creating clickbait and sincere attempts by reporters to present their story in a way that will make it appeal to the audience. We would argue that identifying the bit of your story that packs the most punch is a core journalistic skill that has been around as long as the profession itself. Is there any shame in applying this skill appropriately when meeting your audience in their digital environments? Is the issue here a reluctance on behalf of journalists from 'quality' legacy news organisations to adapt to the reality of social storytelling? So, avoid writing headlines that give a misleading impression of the content and its value (i.e. clickbait), but do make the most of your investigation by selling it in a way that does justice to the importance of your story.

References and recommended reading

Bean, A. 2021. *Crafting a video content strategy for social platforms*. [Online]. [Accessed 16 June 2022]. Available from: <https://www.poynter.org/reporting-editing/2021/video-content-strategy-social-platforms/>

Bucknell I. 2020. Going digital, not dying out: how universities are uniquely placed to teach digital journalism survival skills. Journalism Education. 9 (1), pp. 6-17.

Denisova, A. 2022. Viral journalism. Strategy, tactics and limitations of the fast spread of content on social media: Case study of the United Kingdom quality publications. *International journal of communication.* **16.**

Financial Times. 2024. The Financial Times's profile. [Instagram]. [Accessed 28 November 2024]. Available from: <https://www.instagram.com/p/DCCciOUPO6x/>

Mina, A. 2017. *Memes and Visuals Come to the Fore*. [Online]. [Accessed 17 July 2024]. Available from: <https://www.niemanlab.org/2017/12/memes-and-visuals-come-to-the-fore/>.

Newman, N., Fletcher, R., Robertson, C-T., Eddy, K. and Nielsen, R. K. 2022. *Reuters Institute Digital News Report 2022.* Oxford: Reuters Institute for the Study of Journalism.

Vázquez-Herrero, J., Negreira-Rey, M-C. and López-García, X. 2022. Mind the Gap! Journalism on Social Media and News Consumption Among Young Audiences. *International Journal of Communication.* **16**, pp. 3822–3842.

Online

The internet, designed for people to share information easily and quickly, seems the ideal medium for journalists who want to share news stories easily and quickly. The first online newspapers were distributed as early as 1980 when an experiment by CompuServe and the Associated Press made the Columbus Dispatch available for reading over a dial-up service. It's the invention of the World Wide Web in 1989 that made news more accessible – allowing information to be posted on 'websites' that could be easily searched for and viewed from any computer.

The first online news was text based. However, as technology and bandwidths improved, and access to computers and eventually smart mobile devices grew, so too did the variety of content available. Rather than relying on one medium (print, TV, radio) the advent of news websites means that for the first time journalists have the 'unique opportunity' to combine text, pictures, video and audio all together on one platform to tell their stories (Hargreaves, 2003, p. 254). While many news organisations will publish material on other platforms (such as X, Facebook, TikTok, etc.) their own website (or own app version of their website content) remains the official 'face' of the organisation and is often where audiences are directed to read the full version of a story.

Multimedia journalism can be described (Deuze, 2004) as:

> Presentation of a news story package on a website using two or more media formats, such as (but not limited to) spoken and written word, music, moving and still images, graphic animations, including interactive and hypertextual elements.

In this chapter, we'll consider the benefits and challenges that this multimedia approach can bring. You'll find out more about the diverse types of media that can be incorporated into a web page and how they can be best used to enhance your story. We'll also explore some of the different style of web pages used to tell news stories such as listicles and live blogs. There will be practical and critical exercises throughout to explore the different skills.

Writing for online

How does writing for online differ from traditional news writing for print or broadcast? The truth is that it lies somewhere between the two. Using the structure of the first, but the more informal, simple language of the latter. While many of the journalism and writing skills remain the same, there are a few considerations that are specific to creating content for digital platforms.

Understanding the online audience

Unlike print, where you could usually assume a local population would be reading your work, a news article written for a website or social media can be read by anyone, anywhere in the world. This means simple language is easier for people who may not be reading in their native language. Locating your story geographically can be more important to help the reader understand the context as cultural, financial and legal aspects vary across different areas. You also cannot assume knowledge – if you were writing for a local newspaper you can expect a reader to recognise a local celebrity, politician or landmark but once the story goes online your reader may not have this understanding. So, you need to make sure you're providing necessary details. Puns, sarcasm, dialect and slang should be used sparingly as they may not translate well across global audiences.

Another key to understanding the online audience is considering how they'll be reading your work. What device are they using and where are they when they are reading? While some readers may be viewing on a full-size computer or laptop screen, many will be using a small screen like a tablet or mobile phone. They may be on a commute, sitting in a cafe or scrolling while watching TV at the same time. This reinforces the need for your news story to be presented in a way that is easy to read and attractive enough to compete for the reader's attention.

Let's take this understanding of the online audience and put it into practice with writing text. In this chapter we'll focus on writing news articles for the web, but many of the principles transfer to writing for social media and other digital purposes.

Structure of an online news article

We can begin by looking at the basic narrative structure of your online news story. Traditional storytelling may rely on a linear, chronological structure or a build to a reveal or the weaving of multiple strands to engage the reader.

When writing for online, we use a structure that was first used for news writing in newspapers. That is known as the 'inverted pyramid' structure. This structure is designed to give the reader the most important facts about the story up front and then add in further details and background later.

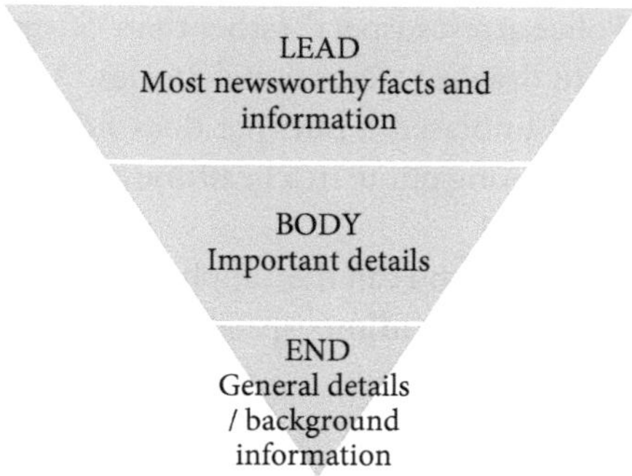

Figure 3.1: The inverted pyramid structure for news writing.

Why is this structure particularly important for online news writing? Research carried out over many years by the Nielson Norman Group has been tracking how people's eyes scan articles online when they are reading. Their research shows that people don't read digital content completely, they tend to scan words rather than take them in order (Moran, 2020). They have identified a number of patterns to people's online reading – similar on both on computer and mobile devices. These include a focus on headings and subheadings first ('layer cake') or a focus on the beginning and (for readers from countries that read left to right) the left-hand side of the screen ('F pattern').

This research helps us to understand how our readers are consuming the content that we write – and show we need to provide enough information in the areas they are likely to scan first to encourage them to read the full story. It's clear that the headline and the opening paragraph are the places in an article that a reader looks for first. So, this is where you need to concentrate your efforts.

Writing effective headlines

The general advice for writing headlines is similar to that you may already be familiar with in traditional print journalism.

- Get attention: Make people click in the first place.
- Summarise the story: Tell your reader what the story is about so they know they're interested.

- Short and snappy: The possibility that the headline will be viewed on a mobile device with a small screen determines the need for a short headline.
- Tell the truth: Your headline should not be misleading.
- Present tense: E.g., 'Police arrest suspect' rather than 'Police have arrested a suspect'.
- Active voice: E.g. 'Police arrest suspect', rather than 'Suspect arrested by police'.
- Emotion. Readers are drawn to emotional stories. As a journalist you may aim to be neutral and unbiased – but that does not mean our interviewees are. The inclusion of a strong quote in a headline can make it more appealing.

With these guidelines in mind, you can then apply what we know about the digital environment to create effective headlines specifically for use online.

Keywords and SEO

It's vital that among the few words you choose for your online headline you have identified 'key words' for your story. To understand the importance of this we need to think about how the world wide web works. Some people will have one or more news websites or apps that they visit and read regularly to stay up to date. However, many more will turn to search engines to look for information on a subject they are interested in.

It's hard to find definitive numbers for the size of the internet. According to the Netcraft Web Server Survey there were over one billion websites on the world wide web (1,104,170,084 to be precise) in July 2024. One online tracker, WorldWideWebSize, is updated daily and suggests there are at least 3.29 billion individual pages in the indexed web. Of course, many of these pages may no longer be updated and are classed as inactive. New sites are created and removed every day.

Whatever the numbers, I think we can agree the amount of information available online is vast. To help us navigate this huge amount of data and find what we want we rely on information retrieval software known as search engines. When you type in a search term the search engine uses an algorithm to look for the most relevant results. It does this by searching for key words in the text or the meta data of your web page. Meta data is data that provides information about other data. In the case of a webpage, there can be descriptions, tags and key words that are hidden in the background code for the page, rather than displayed to the reader. The process of maximising visitors to your website by using words and structures that work with algorithms is called Search Engine Optimisation (SEO).

If you want a reader to have the best chance of finding your story you need to think about the words they would use to search for that story and ensure they are included in your writing. These are the 'key words' for your story. The algorithm will rate your story more highly in the results if these key words appear early on (i.e. in the headline and first paragraph).

Taking the time to choose these key words carefully can make a dramatic difference in how easy it is for readers to find your story. Some factors are out of your control (for example, search engines will generally rank trusted sources more highly, so if you work for a well-known news organisation you'll have an automatic advantage). However, it's worth taking time to consider what the audience might search for while crafting your headline. Here are some general tips to help.

- Be as specific as you can. For example, 'Royal couple visit UK city' is less likely to be searched for than 'William and Kate visit Leeds'.
- If it's an ongoing story make sure you use the obvious key word in the headline. This is often done on news websites by using the same word or phrase for every article on the subject, followed by punctuation and then the headline for the individual story e.g., 'Election 2024: Record turnout at polls' 'Paris Olympics: Second gold for Team GB'.
- If there's something relevant that people are talking about, include it in the headline. Think about how people refer to a story not necessarily the technically accurate term (as long as it's still correct, you can use the full term later in your article). For example, people are more likely to search for the term 'Brexit' rather than 'withdrawal of the United Kingdom from the European Union' or 'Covid' rather than 'SARS-CoV-2 virus'. You could use online services such as Google Trends or Trending on X to see what words, phrases or hashtags are currently in popular use about your topic.
- If you're linking to another story make sure that the source is recent and trustworthy. Links between pages can impact how search engines rate your story.

Search engine algorithms are constantly being tweaked so it's worth trying to keep up to date with what works. We'd encourage you to, well, use a search engine to find some of the many resources available online (some provided by the search engine companies themselves) to help you understand and improve your SEO knowledge.

Clickbait headlines

Advertising expert David Ogilvy famously wrote in a marketing advert that ran in the USA in 1970s, 'On average five times as many people read the headline as read the body copy. It follows that if you don't sell your product in the headline, you've wasted 80 % of your money.' In the online world where the number of readers clicking on your articles can lead to financial and reputational benefit, this is as true of selling your story to your reader as it is of selling products. Be wary of trying to play the algorithms at the expense of your journalism and editorial judgement though. First and foremost a headline needs to make sense and to describe your story, then you can consider how to do this in a way that will attract the reader.

As we've seen, the art of writing a headline for use online is different to in print. That may be due to screen space, the use of keywords or other tactics used to attract attention. However, headlines that have been specifically designed to attract the attention of viewers in this way can have a downside. Internet users are aware of and savvy to 'clickbait' tactics. Example of these tactics include the use of headlines that sensationalise or exaggerate the story, or those that only tell you part of the story (e.g. 'You'll never guess what this woman did next'; 'The common food that scientists claim could be bad for your health'; 'Celebrity reacts to those controversial photos'). These headlines appeal to readers by piquing curiosity or referring to topics (often health, finance or relationships) that make it seem the story will be impactful to people's personal lives.

Different news organisations have different approaches to the use of clickbait style headlines, with digital-first news sources often being more open to them than more traditional print or broadcasters. In general, although by no means exclusively, these kinds of headlines are more often applied to lifestyle content than articles about harder news such as politics.

Although using these tactics can entice more people to click, there are concerns about using them for news articles. Readers can feel disappointed or misled if there's a disconnect between the exaggerated headline and a story that is not quite as suggested. Research has found that the use of these tactics in headlines can also result in the reader feeling that the news source is less credible (Molyneux and Coddington, 2019). If you're going to consider using clickbait tactics, make sure the article's content delivers the story the headline promises.

Exercise 3.1: Practice your headline writing and test them out

There are a number of free AI headline analysers available online. Google 'headline analyser' (the US spelling works best because many are made by companies in America) to find a selection.

You can type a headline into one of these to get a report on how effective it might be. Most of the analysers tend to look at word count, language and SEO as well as concepts such as emotional appeal. Try a headline from an online news story of your choice. Could you rewrite the headline to improve the score? Does it still make sense?

> Remember that these AI tools are computer programmes and each have their own biases and priorities. You can use them as tools to help you think critically about your headline writing but don't think of them as infallible experts. Always use the headline that you believe best tells your story.

Summary paragraph

According to research on reader traffic and engagement (Chartbeat, 2020), 45 % of readers who load an article will leave within the first 15 seconds. The average time spent on an article is 30 seconds. At an average reading speed of 200 words a minute – that's fifty words in 15 seconds! Almost half the people clicking on your page will only read the first fifty words. So, it's vital to make them count.

Your first paragraph of your online news article should summarise the story – use this space to reinforce and expand on your headline but don't just repeat it. Sometimes this first paragraph may be the copy that appears under the headline on a homepage, so it needs to explain what the reader needs to know about the story and make them interested enough to click and read further.

We can rely on our traditional journalism writing skills for this, with some tweaks for the online audience.

- WHAT? Think about what has actually happened! Press releases can focus on one thing, but is that the main story?
- WHEN? Unlike newspapers, where today's news is famously 'tomorrow's fish paper', a digital story lasts forever, and your article may be seen at any time. So, avoid phrases like today, yesterday, last week, an hour ago. Instead give days, dates and times. You can create a timeline by referencing relevant milestones in the story, 'this comes a week after…'
- WHO? It's important to get names and titles correct in a medium where it's possible to search for more information about anyone.
- WHERE? Remember your audience may not be based in the same place as you. As a rule you would start with a reference to a wider geographical area and then get more specific further on in the article. For example, 'London' first, then 'Islington', then if relevant you can get down to more detail such as street names.
- WHY AND HOW? This background (which may include links, explanation, interviews) can be expanded on in the rest of the text.

Exercise 3.2: Now write both a headline and summary of a story

One of the first digital news services was Ceefax (see historical context in introduction). Launched in September 1974, Ceefax was a system that delivered text (news, sport and more) that BBC viewers could access on their TV screen, using their remote control to select pages.

The system for writing these pages was very restricted. The template to fit on a TV screen allowed for just a headline (of exactly thirty-three characters – including spaces). Text was limited to four paragraphs, with a maximum of eighty words in total.

Although the Ceefax service finished in 2012, the importance of writing succinctly still applies today. Many readers use mobile phones where a headline and short summary is still all that can initially be seen on the small screen.

Practice telling a story in a short format. Use the Ceefax rules to write a headline and summary for the plot of a film/tv show – can your friends guess what it is?

Writing style

In the online world, as in the traditional media, there are no fixed rules for writing style. Depending on your publication style, your intended audience and the type of content you're producing, you may have different styles. However, there are some best practice guidelines for news articles, based on what we know about how people read news online.

- Use active rather than passive sentences, as mentioned for headlines.
- Write in the third person not the first person.
- In online news writing, a paragraph is usually one sentence long.
- Try and stick to one idea per short sentence, avoiding lots of clauses.
- Avoid technical language and jargon. News is aimed at a wide audience, e.g., avoid 'primary care services' and say 'doctor' or 'GP'.
- Depending on your story, it's possible to write in a more informal style online, where readers tend to be from a younger demographic than print. Write in a way your audience can relate to.
- Remember your basics – spelling and grammar. It's even more important to get this right in a medium where your story can be archived forever, and your mistakes can be noticed and commented on by a vast number of potential readers.
- Did you know that most news organisations have their own online style guide? It's worth looking at them to see how they approach writing for different platforms. Examples include the BBC news online style guide, the AP style guide and the New York Times stylebook.

Exercise 3.3: Write a complete online news story

Use a press release from a local council/government/emergency services press office to write an online news story using the conventions above.

Remember to use the inverted pyramid structure and to use the correct writing style for online news.

Multimedia web pages

What is a multimedia web page?

Multimedia means 'using more than one medium of communication or expression'. In the context of an online news web page, multimedia elements can include, but are not limited to:

- Body text
- Additional textual elements
- Pictures
- Video
- Audio
- Hyperlinks – reader clicks to get more information
- Interactive elements – these allow the reader to do actively do something to engage further with the story

News organisations and websites all have their own style guides, templates and publishing software, so few journalists will be expected to create a web page from scratch using all aspects of multimedia. However, the process of considering how different elements can complement text and tell a story is both a stimulating intellectual exercise and develops production skills.

This article from *The Guardian* (Readfearn, G., 2024) in *figure 3.2* looks at the state of the coral reefs in Australia. The text alone tells the complete story in the same way as it would in a printed version of the article, allowing the reader to quickly get the news information they need. However, this online article also includes several simple multimedia elements to give the reader a richer experience, with opportunities to delve deeper into the topic and to engage further if they are interested.

Multimedia elements used in this article include hyperlinks to further information on both internal (within the same website) and external (on other websites) articles – including a picture promotion to another related article in the

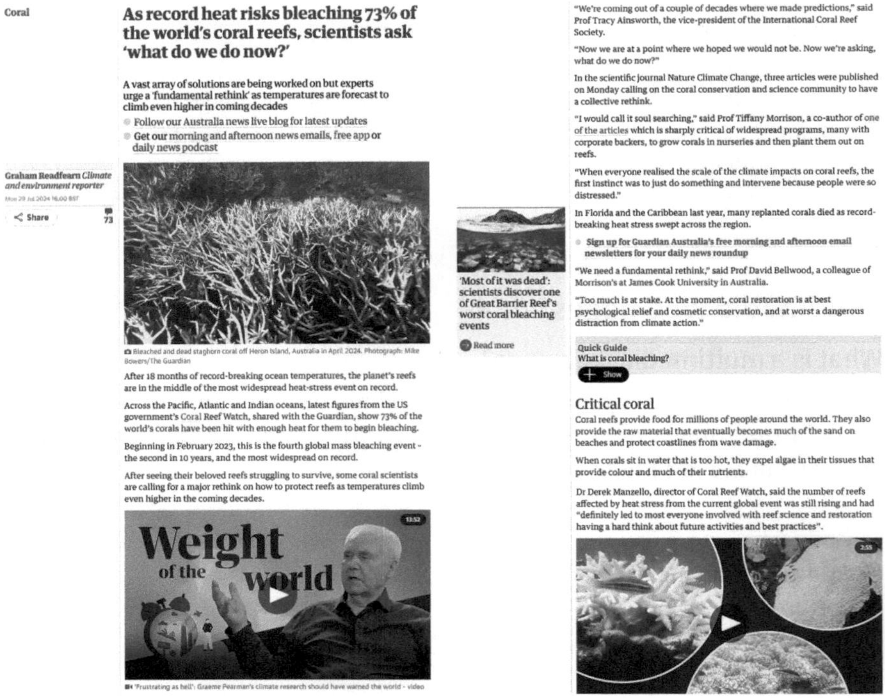

Coral

As record heat risks bleaching 73% of the world's coral reefs, scientists ask 'what do we do now?'

A vast array of solutions are being worked on but experts urge a 'fundamental rethink' as temperatures are forecast to climb even higher in coming decades

- Follow our Australia news live blog for latest updates
- Get our morning and afternoon news emails, free app or daily news podcast

Graham Readfearn *Climate and environment reporter*
Mon 29 Jul 2024 16.00 BST

Share 73

Bleached and dead staghorn coral off Heron Island, Australia in April 2024. Photograph: Mike Bowers/The Guardian

After 18 months of record-breaking ocean temperatures, the planet's reefs are in the middle of the most widespread heat-stress event on record.

Across the Pacific, Atlantic and Indian oceans, latest figures from the US government's Coral Reef Watch, shared with the Guardian, show 73% of the world's corals have been hit with enough heat for them to begin bleaching.

Beginning in February 2023, this is the fourth global mass bleaching event - the second in 10 years, and the most widespread on record.

After seeing their beloved reefs struggling to survive, some coral scientists are calling for a major rethink on how to protect reefs as temperatures climb even higher in the coming decades.

Weight of the world

13:52

'Frustrating as hell': Graeme Pearman's climate research should have warned the world - video

'Most of it was dead': scientists discover one of Great Barrier Reef's worst coral bleaching events

Read more

"We're coming out of a couple of decades where we made predictions," said Prof Tracy Ainsworth, the vice-president of the International Coral Reef Society.

"Now we are at a point where we hoped we would not be. Now we're asking, what do we do now?"

In the scientific journal Nature Climate Change, three articles were published on Monday calling on the coral conservation and science community to have a collective rethink.

"I would call it soul searching," said Prof Tiffany Morrison, a co-author of one of the articles which is sharply critical of widespread programs, many with corporate backers, to grow corals in nurseries and then plant them out on reefs.

"When everyone realised the scale of the climate impacts on coral reefs, the first instinct was to just do something and intervene because people were so distressed."

In Florida and the Caribbean last year, many replanted corals died as record-breaking heat stress swept across the region.

- Sign up for Guardian Australia's free morning and afternoon email newsletters for your daily news roundup

"We need a fundamental rethink," said Prof David Bellwood, a colleague of Morrison's at James Cook University in Australia.

"Too much is at stake. At the moment, coral restoration is at best psychological relief and cosmetic conservation, and at worst a dangerous distraction from climate action."

Quick Guide
What is coral bleaching?
Show

Critical coral

Coral reefs provide food for millions of people around the world. They also provide the raw material that eventually becomes much of the sand on beaches and protect coastlines from wave damage.

When corals sit in water that is too hot, they expel algae in their tissues that provide colour and much of their nutrients.

Dr Derek Manzello, director of Coral Reef Watch, said the number of reefs affected by heat stress from the current global event was still rising and had "definitely led to most everyone involved with reef science and restoration having a hard think about future activities and best practices".

2:55

Figure 3.2: A screen grab showing part of an article from The Guardian about coral bleaching.

left-hand column of the page. There is a photograph that clearly reflects the story, used to add visual interest. Two embedded videos – one allows the reader to see an extended interview with someone quoted in the article itself and another gives additional background information on the topic. There is a 'Quick Guide' explainer where the reader can click to see extra textual context. There is also an interactive section for readers to engage further with the story by leaving their comments. The reader could get the information they need by reading the text, but if it's a topic that interests them the multimedia elements allow them to access additional details, gain further understanding, explore related topics and engage with other readers and the journalist who produced the story to share opinions.

Other articles may use more sophisticated and complex multimedia elements to add interactivity to a story, such as including data that can be interrogated and searched by the reader in real time. Examples could include a database allowing users to check the election results in their area, an interactive map showing coastal erosion over time, or a timeline that users can click on to find out more detail on how a news event unfolded.

Elements of a multimedia webpage

It's easy to be overwhelmed by the constant changes and developments in different platforms and technologies involved in digital journalism. The good news is that the journalist's role is not necessarily to know the computer coding, software and technical capabilities – but rather to understand how they apply to the telling of stories. With multimedia online news stories, that means selecting interactive elements such as video, audio, polls and hyperlinks to enrich the reader's experience and present information in an engaging, immersive and accessible way.

The journalist working on a multimedia online report has many options available to them to help tell a story. A news article on a website could be told effectively with only text and pictures, but engagement could be increased by adding a video, a fact box or a link to another source of information. It's vital to consider the purpose of including a multimedia element as the reader does not click on the link or watch the video just because it's there. They choose to click because they believe it will enhance their experience – it will interest them, help them understand the story, or bring a new perspective. Multimedia for multimedia's sake will not provide this.

Consider the potential value and drawbacks of using each type of media as part of a web page:

Table 3.1: Elements of a multimedia webpage

Multimedia element	Examples	Potential value	Potential drawbacks
Text	Main text of the article.	Presents the story.	Can be too long.
Additional textual elements	Sub heading. Fact box / side bar. Bullet point list.	Breaks up text, draw the reader's eye. Helps present additional information, background or statistics in an interesting and simple way.	Overuse can be confusing for the reader.
Video	Social shorts. Interview clips. Animation. Gifs. Eyewitness footage.	Presents the story in a visual way. Allows reader to see contributors (case studies, experts, etc.) first-hand. Can include extended interviews. Aids further exploration of subject.	Reader can decide whether or not to click on and watch these videos. If there's something it's essential for the reader to know, it must also be included in the text. Reader may be using technology that does not allow video watching (e.g. limited data, small screen). Reader may have accessibility issues that mean they cannot view video.

(Continued)

Table 3.1: (Continued)

Multimedia element	Examples	Potential value	Potential drawbacks
Audio	Interview clips. Voxes. Packages. Podcasts.	Presents the story in an audio format. Allows reader to hear contributors (case studies, experts, etc.) first-hand. Can include extended interviews. Aids further exploration of subject.	Reader can decide whether or not to click on and listen to audio. If there's something it's essential for the reader to know, it must also be included in the text. Reader may be using technology that does not allow audio listening (e.g. limited data), or may have the sound turned off on their device. Reader may have accessibility issues that mean they cannot hear audio.
Pictures	Photographs. Infographics. Maps. Photo galleries.	Presents the story in a visual format. An attractive picture will catch the reader's eye. Presents complicated information or statistics in an interesting and simple way.	Using a picture without a strong connection to the story may be confusing for the reader. Reader may have accessibility issues that mean they cannot see visual elements.
Hyperlinks	Internal – to other articles within the same website. External – to other websites.	Allows further exploration. Shows readers different angles to a story. Points to primary or additional sources of information.	When the reader clicks on a link, they leave your article.
Interactive elements	Online polls. Comments or message boards. Interactive map. Searchable database.	Allows further exploration. Encourages reader to take an active part in the story. Gathers responses and reaction from readers about your subject.	Can be complicated and timely to produce. Comments and messages will need to be moderated.

As well as considering which multimedia elements you could use, it's important to make sure they're being used effectively. Excessive use of multimedia could distract, divert or confuse the reader and detract from the story the journalist is trying to tell. As Nielsen and Loranger note, it's important to 'use multimedia thoughtfully, and only when it adds substantial value' (2006, p. 375). Pictures and videos should illustrate the story as closely as possible. Using an image will draw attention to a story, though use of stock images or disconnected images can be confusing for the reader. Consider whether hyperlinks will take the reader to information that they'll find interesting because it illuminates an idea touched upon in your article.

Connect to primary sources where possible, avoid promoting rival publications and remember that if they click on the link, they'll leave your story.

Storytelling

As well as ensuring that each multimedia element is meaningful, Jim Hall says that the author must think through how they fit together to form a coherent report:

> What is important for the journalist is the ability to conceptualise a piece of journalism in multimedia and interactive terms. The textual element of the journalism no longer... carries the story. There will be some information for which it is the appropriate form, but in order to make its effective contribution to the story it needs to be organised alongside, or in sequence with, other pieces of information (lexias) for which different forms are more appropriate. (2001, p. 91)

So, when producing a multimedia online news story, the digital journalist must consider the various aspects of the story and match them to the most appropriate medium to convey that information. Medvedeva (2016) suggests the use of morphological analysis for this purpose. This is a brainstorming technique that 'involves conceiving of two or more independent dimensions of a problem, coming up with a number of attributes for each dimension, and then examining all the combinations' (Shoemaker et al., 2004, pp. 151–152). Using this methodology, Medvedeva shows how, as an educational tool, a student can consider two dimensions of a multimedia project at the same time – the content and forms of the content. Using a similar process, a multimedia planner used by a digital journalist can show the different elements of the story and the media used to illustrate these elements. It can also help them to consider the journey of the reader through the page.

Multimedia planner

Table 3.2: A planning grid for your multimedia productions

	Project name					
Multimedia elements	**The main narrative**	**Contributors**	**Events**	**Statistics**	**Background**	**Exploration**
Text (500–600 words)						
Additional textual elements						
Photograph(s) and images						
Video						
Audio						
Hyperlinks						
Interactive feature						

Here is an example of the planner in practise, as applied to a story from the Daily Mail (Tomkin, 2022) about the first pictures to be released from a NASA space telescope.

Table 3.3: A completed multimedia planning grid

	NASA unveils the first EVER images from its James Webb Space Telescope including a selfie ...					
Multimedia elements	**The main narrative**	**Contributors**	**Events**	**Statistics**	**Background / explainer**	**Exploration**
Text (500–600 words)	First pictures from NASA's James Webb Space Telescope	Professor of astronomy, University of Arizona. Astronomer from the Space Telescope Science Institute.		Cost of telescope, main facts about the telescope.	Separate section 'The James Webb Telescope'.	
Additional textual elements				'Facts & Figures' sidebar fact box.	'Instruments on the telescope' sidebar fact box.	
Photograph(s) and images	Gallery of pictures from the telescope.				Infographic of the telescope. Map of location.	
Video	NASA video about the telescope.					
Hyperlinks					Internal links in text to previous articles about the telescope.	Keyword buttons at bottom of text to link to other stories on 'NASA' and 'Space'.
Interactive feature						Buttons to share article with other people. Comments section.

Remember that your key journalism skills, news finding, interviewing, writing and research are what makes a successful story. The multimedia elements are simply embellishments to help you tell that story in an engaging way.

Exercise 3.4: Analyse a multimedia news story to understand its characteristics

Use what has been discussed so far to analyse a published news article. Select a recent web page from an established news website.

What multimedia elements have been included in the telling of the story? Use the multimedia planner to organise your findings.

Can you identify what value each element adds for the reader of the story?

Do you think the use of multimedia has been relevant – is there anything you would change?

Production

Many websites are created using Content Management Systems (CMS). These systems allow users with even minimal technical ability to be able to input content and information, which the software then adds to templates and presents as web pages. This means the journalist can think about their story rather than having to understand or master the computer coding and technology needed to create a website. Some media organisations use bespoke in-house systems, some use more widely available solutions. They may differ in how they work and look, but the principles are the same.

For this book we'll look at WordPress. In June 2022, WordPress was the most common Content Management System, used by 29.8 % of the top 1 million websites (builtwith.com, 2022). According to the WordPress website (January 2025), news organisations that use the system for their websites include the Time magazine, Al Jazeera, USA Today and News Corp. WordPress is also a free and open-source software making it easily accessible. So that is the software we have used to illustrate the examples in this book. While the details may change the broader principles remain the same whatever CMS system is being used. For example, you could explore other website building platforms that offer free access such as Weebly and Wix.

Using the block editor in WordPress allows the digital journalist to consider each multimedia element as a different block. You can switch to block editor from the WordPress dashboard under Settings > Writing. Once you have started a post, you can access the block menu by clicking on the plus symbol in the editor, highlighted in red below. You can search for the block by name if you know the one you want, browse all blocks by category or select from the shortlist of frequently used blocks.

Figure 3.3: A Word Press CMS screengrab.

Most blocks are obvious – you can add separate paragraphs of text, headings, pictures etc to create the basic article. Investigate the system further though, as there is a selection of block formatting options that you can use to add interest and create multimedia webpages to help your reader explore your story in more detail.

- **Accordion:** Prompts readers to click on an arrow to reveal extra text.
- **Divider:** Creates a dividing line on the page which can help to separate different parts of the story. There are adjustable settings in the pane on the right of the screen to alter the colour and thickness of line. The Spacer block is similar but with different formatting options.
- **Container:** Puts a block of background colour around a chunk of text, or a piece of media, or a combination of the two.
- **Shortcode:** This allows you to embed an *iframe* code into your webpage to include more interactive elements (as an example see Google Maps section below).

There are many more options available that could enhance online production. These blocks will automatically create design or interactive features in a page without the journalist needing to know the coding to create them. This type of editing software is known as WYSYWIG (What You See Is What You Get). Most CMS software is like this and there will be equivalent ways to add features such as menus, dividers, text boxes and embedded elements as those shown above for Word Press. Some CMS systems will be more restrictive and do not allow the journalist as much autonomy over design, instead asking them to input text into a ready created template.

As well as writing your story and considering how it will be presented, it's important to look at other hidden elements of your web page. Adding 'alt text' and 'alt descriptions' on visual elements will improve the accessibility of your website. Alt text is used in place of your image if for some reason the picture won't load properly on someone's screen. Alt description will be used by screen reading software to describe your picture to someone who has a visual impairment and can't see it. Look out for any fields in your CMS with 'description', 'key words' or 'meta tags'. These will add to the meta data of your page and provide SEO information that can be used by search engines to find and rank your page.

Interactivity

There are many online tools available to easily add interactive elements to a website, often for free. Here we have included just a few examples.

Online polls

If you want to gauge quick reaction to an element of your story, a poll offers your readers the chance to vote on a question by choosing between a few set options. There are lots of tools available to create a poll for a webpage, such as the free Crowd Signal.

Interactive maps

Google's My Maps is a useful tool for adding interactive maps. It allows you to make a map with pop-up information and generates a URL that you then embed into your webpage.

Quizzes

A search for 'free online quiz maker' will give you a few websites that allow you to build an interactive quiz to complement the topic of an article. Once you have completed the options, you ultimately receive a chunk of code to copy into your CMS text editor.

Comments

Many CMS systems will have a built-in option for a comments section. There are a number of plugins available to customise comments sections in Word Press, for example.

Exercise 3.5: Make a multimedia online news story

Create your own multimedia web page in WordPress (or a similar system). Plan your different elements using the multimedia planner, thinking about what value each element brings to your story. Include text, pictures, audio or video, hyperlinks and an interactive element. Experiment with the different blocks available via the block menu, and their settings to add multimedia and design elements to your page.

You can refer to other chapters in this book for tips on creating different types of content for the page.

As well as the multimedia news article, there are different writing formats that have evolved in digital journalism to make the most of the online environment. We'll look in more depth at a couple of these now.

Live blogs

What is a live blog?

Live blogs are web pages providing a commentary on an event while it takes place, typically in the form of frequent short updates. The format has been described as an 'established part of the news ecology, representing a format that is native to the web' (Thorsen and Jackson, 2018, p. 3).

The main characteristics of a live blog are:

> It is live. The page is updated regularly with bite-sized bits of content as they happen. The information is arranged with the most recent content at the top of the page, allowing the user to get the most recent updates first and scroll down to see how the story developed. Although a blog receives most views as it's being updated live, readers do return to them as an historical, archived account of how an event unfolded (Thurman and Walters, 2013).

The tone is conversational, open and honest. Reporters will admit if they have got something wrong and update the reader if they are still waiting for information. They will also reflect reader comments and ask for interaction. Tereszkiewicz says:

> The discourse of news live blogs is the discourse of intertextuality, interaction and dialogue … Informality and stance markers strengthen interactivity, which, together with an emotional charge, make up the extraordinary tone of the blogs and distinguish live reports from the reporting formats used online so far. (2014)

It can include many voices. Different reporters can update the blog, and it can include analysis, quotes and reports from a variety of people all in one page.

A live blog often draws from material from outside sources such as social media and other news organisations. In this respect it combines conventional reporting with curation (Thurman and Walters, 2013), where journalists sift and prioritise information from a variety of places and present it to the audience in close to real time.

It utilises a wide range of multimedia elements – Thurman and Schapals (2017, p. 286) found that live blogs 'contain about fifteen times more multimedia elements than print articles and nearly five and a half times more than traditional online articles', though within the context of being approximately twelve times longer than standard online reports.

Live blogs have proved popular with both news organisations and audiences. Matt Wells, former blogs editor for Guardian online says:

> They provide a useful way of telling stories characterised by incremental developments and multiple layers. They are open about the limitations of journalism and draw in the expertise of the audience – and even take input from journalists on rival publications. On fast-moving stories, live blogs give the ability to post significant developments quickly – more quickly than editing and re-editing a news article. They also allow us to link out to other coverage, to include comments from Twitter and Facebook, to display multimedia (pictures, video and audio), and to include our audience in the comments below the line – all in one place. (2013)

One survey (Thurman and Newman, 2014) showed the widespread popularity of the format – 11 % of UK news consumers had followed a live news page in the previous week, with live news pages even more popular in the US, Brazil, Italy, and Spain, and especially in France (19 %) and Japan (3 %). Another looking at Guardian online content (Thurman and Walters, 2013) found that the median unique visitor numbers for live blogs was 233 % higher than conventional articles and 219 % higher than picture galleries on the same subject. This popularity could be due to several factors. It's easy to use live blogs to keep up with a news event via quick updates while you're doing something else (e.g., at work on your computer, at a social event on your smartphone). It provides one point for readers to come for the latest updates, which will be at the top of the page, rather than having to scroll through long articles. Readers are also encouraged to pay multiple return visits to the page, to keep up with the latest news.

The experience of reading a live blog about a sports match or entertainment event can in some way replicate the communal spirit of attending the event in person. The multitude of voices, perspectives and sources means even if you're attending the event in person you can get a broader view of what is happening by following a live blog at the same time.

There are pitfalls to the format though. In 2011 journalist John Symes referred to the move towards live blogs as the 'death of journalism', saying:

> There is no structure and therefore no sense, and the effect is of being in the middle of a room full of loud, shouty and excitable people all yelling at once with all the phones ringing, the fire alarm going off and a drunken old boy slurring in your ear about 'what it all means'.

Live blogs can run to dozens of pages, quoting so many diverse sources and different media that it can be confusing and hard to follow the narrative thread of the story. A journalist working on a live blog needs to remember that very few readers will see every update and many will join mid-way through the blog and pop in and out. A clear layout, obvious structure and signposting, and use of summaries can help.

The very nature of live blogs means that they rely heavily on a multitude of sources and the fast-paced nature of breaking news means there is often little or no time for verification. There are a number of strategies that digital journalists can use to combat this. In analysis of sources used in live blogs, Tereszkiewicz (2014, p. 308) finds that mainstream media is the most quoted source in blogs – either reporters from the organisation writing the blog or external organisations – followed by government sources and then independent organisations. This can show a variety of opinions but also suggests that credible and attributable sources are chosen more regularly. So, think about where you get the content in your live blogs, verify wherever possible, and make it very clear where the information you're quoting has come from. It's important to be transparent about the process, making it clear to readers that the blog is produced live, and acknowledging and correcting mistakes quickly.

Exercise 3.6: Analyse a live blog as it's being created

Watch a live event such as a sport match or awards ceremony (this only really works if the event is live, so you can't watch on demand) and follow a corresponding live blog at the same time. What did you find out that you didn't know from watching the event yourself? Which bits were most useful/interesting? What was the tone of voice used? Did you notice any mistakes? If so, how were they addressed (if at all)?

Story selection

The kind of stories that work as live blogs are ones that are fast-moving and developing live and that lots of people are talking about. This can include breaking news events such as election results, terrorist attacks or extreme weather events. Local websites may use them to provide regular updates for readers on changing news situations that will affect people in the area such as road closures or building fires. Live blogs are also often used for scheduled events such as sport, entertainment shows or award ceremonies where people may be collectively experiencing an event.

Production

A live blog is usually written and updated by a reporter who is not actually on the scene of the event itself, but rather based in the newsroom. The skill of live blogging is to be able to keep across multiple sources – which could include on the ground reporters, other news sources and social media – and pull together and curate them into one cohesive structure. A live blog can be updated by more than one reporter but in that case it's important to monitor for consistency and to avoid repetition.

There are certain techniques that you'll see used in the production of live blogs, many of which you can see illustrated in the examples included in *figure 3.4*.

- Reverse chronological: Updates presented with most recent at top of page, i.e. reverse chronological order.
- Time stamps: Updates include the publication time, so readers can see how long it has been since an update and get a feel for how quickly things are happening.
- Bite-sized content: Usually one or two short paragraphs of text per post. There is often a visual element (such as a line, graphic or obvious space) to separate pieces of content, to make it easier to read.
- Regular summaries: For example 'What we know so far', so readers don't have to scroll through masses of information to get an overview of the situation. Sometimes these summaries are pinned to the top of the page, or a side bar, to make them easy to find.
- Multimedia: Make use of video, photos, maps, graphics.
- Voices: Include analysis and updates from a variety of voices – including reporters, experts, eyewitnesses, etc. These will all be clearly named to show where the information is coming from.
- Embedded social media: Posts from X, YouTube, Instagram, Facebook, etc. can be added to the live blog using their URL or embed code. This is an efficient way to add media and voices.
- Tone: A conversational tone of voice – especially for less heavy topics – is often the way to go.
- Hyperlinks: Point readers to other sources for further information (for example narrative articles, explainers or previous related news stories) with hyperlinks.

Many news organisations use specialist live blogging software to create their pages, or have a similar function within their own content management systems. You can replicate the style of a live blog easily within WordPress by creating a page that is updated regularly, using the production techniques outlined above.

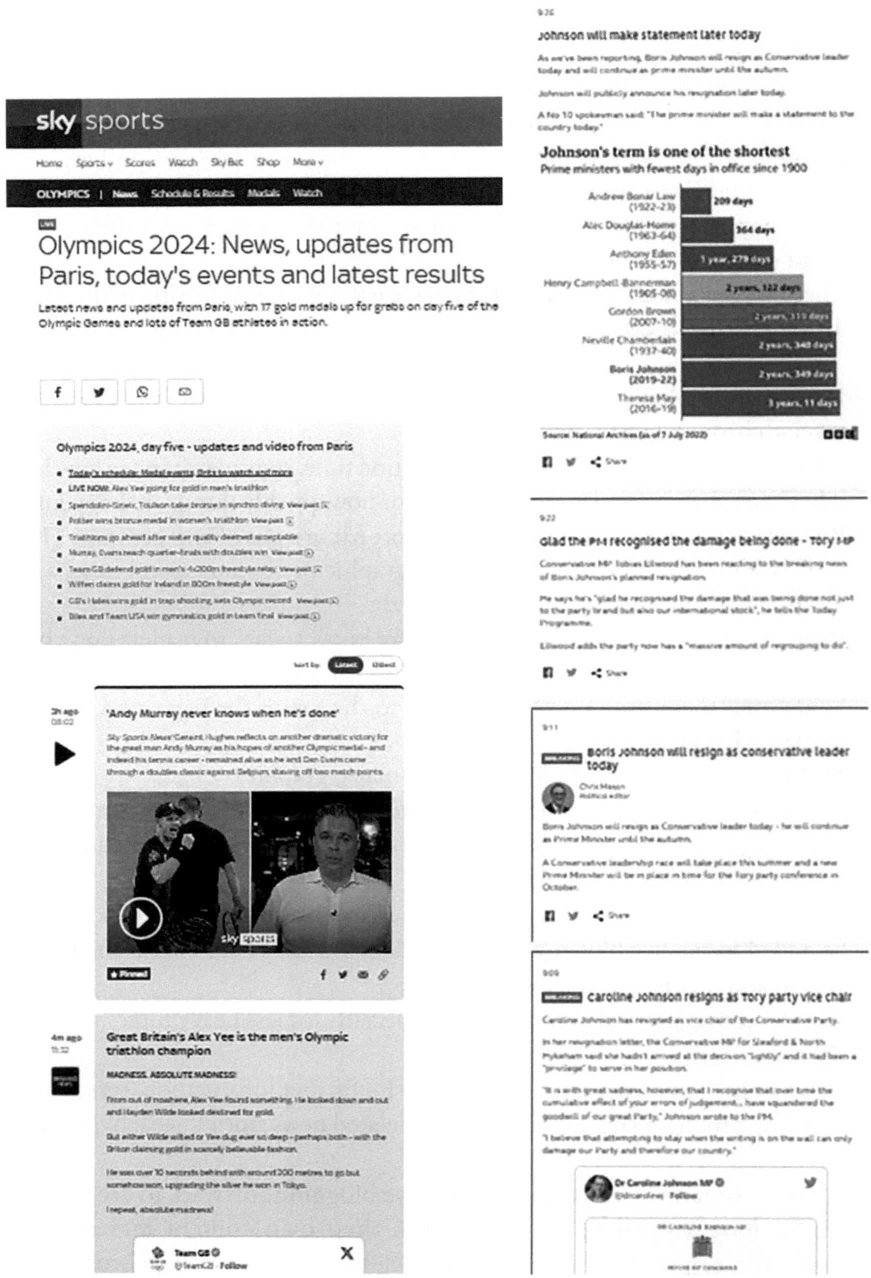

Figure 3.4: Extracts from a Sky Sport live news blog 1 August 2024 and a BBC News live blog 7 July 2022.

Here are some of the block formatting options that you may find useful:

- Use *Divider* or *Spacer* blocks to make an obvious separation between each update – this will make it easier for readers to navigate.
- *Container* – puts a block of background colour around a chunk of text, or a piece of media, or a combination of the two. This can help essential information to stand out.
- Use *Headers* for every update and remember to include the time at the start.
- The *link button* can create hyperlinks to other websites or articles.
- *Testimonial* can create a by-line with a photograph for different reporters.
- Use different *Embeds* blocks to include social media – or simply screengrab and include as pictures.

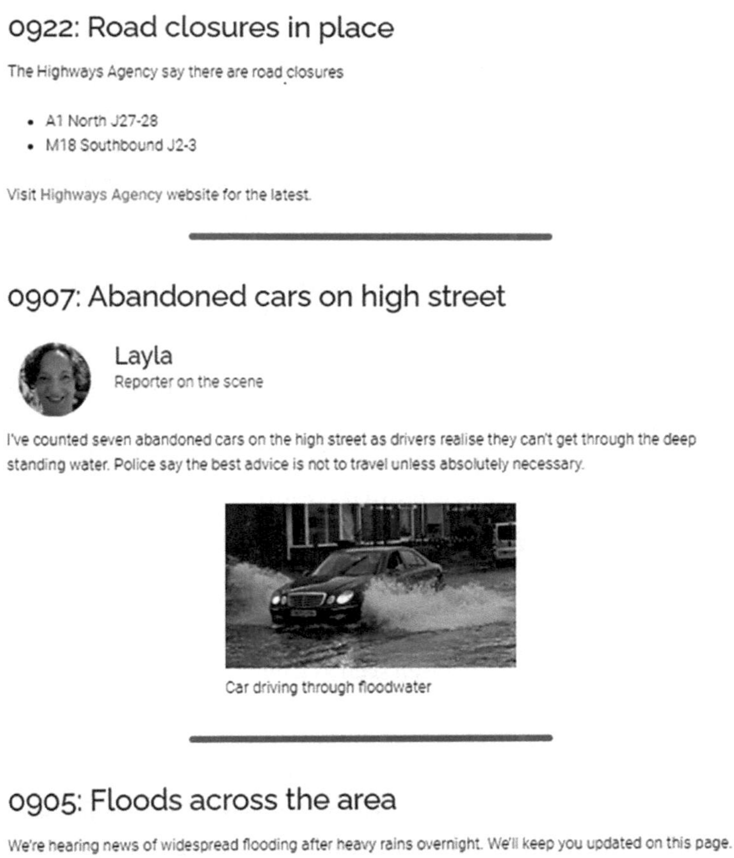

Figure 3.5: An example of a very simple live blog created using WordPress.

Options to create similar formatting blocks will be available in other website editing systems.

> **Exercise 3.7: Write a live blog on a news event of your choosing**
> Create your own live blog using WordPress (or a similar system). Think about the best story to cover using this format to ensure you can get regular updates from multiple sources using a variety of different media.

Listicles

What is a listicle?

A listicle is an article that is written in the form of a numbered list. There will usually be a brief introduction to the topic, then each item on the list is presented with a heading, a picture and a small amount of information. While lists have been used in journalism for a long time, the listicle is a form that really came into its own on digital platforms. It's often associated with digital-native media such as BuzzFeed or Huffington Post but are used by a variety of news organisations.

Listicles can be seen as a lazy form of journalism, filled with repackaged information, empty links and dismissed as just 'click bait' designed to get shared (Holic, 2014). However, a study of US undergraduate students (Sadri, 2019) found that they saw a BuzzFeed listicle as significantly more credible than a traditional USA Today article.

The idea is that breaking a story down into numbered chunks might make it easier to read and remember, and that creating a numbered list encourages the reader to continue reading until they reach the end of the list, to see what's included.

Story selection

In news journalism, listicles are often seen as a light-hearted format used mainly for entertainment and lifestyle stories. But this does not have to be the case. A numbered list can be an effective way to break down complicated information and explain difficult topics and make them easier to understand. Hence it can be used in stories where there is a lot of data involved. A listicle is also good for anything where there is naturally an element of ranking – this could be sport, consumer deals, highest crime rates, etc.

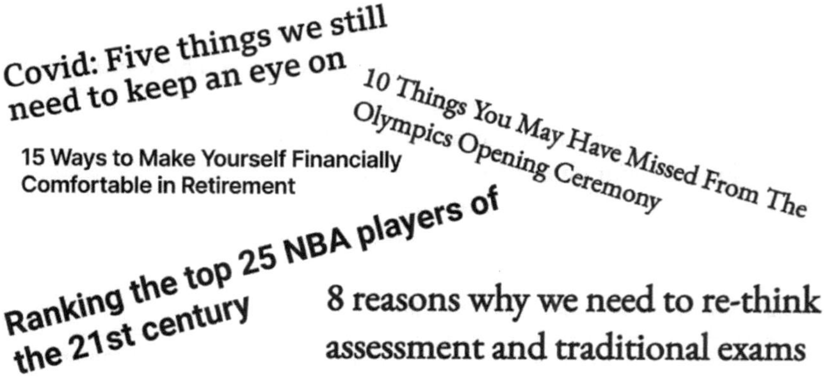

Figure 3.6: Some examples of listicle headlines.

Production

Is there a perfect number of items to have in your list? In 1956 a psychologist called George Miller suggested that there's a limit to human capacity and that best understanding could be found at the 'magical number seven' (plus or minus two). This would suggest that the ideal list should be between five and nine items long. But there is not really any perfect number. Let the story lead you – the most important thing is how many items it will break into, not stretching it out too thin, or cutting things out to reach a certain number. That said, a too short list would seem pointless and a too long list would be off-putting for the reader.

There are some techniques that you'll see in the production of news listicles

- Start the article with an introduction that includes the main news purpose of the list. This sets the scene and tone for the list and anchors it in news rather than making it seem frivolous.
- Remember to include the numbers of each item on the list in the article – it helps the reader to navigate.
- Consider the narrative structure. The items should all lead on from one another and tell one cohesive story.
- If your list is in order of merit, it's conventional to write as a countdown and finish with the number 1 item. This will encourage readers to continue to the end to see who is top of the list.
- Keep a consistent format for each item on this list. Don't have some that are longer than others or contain different information.
- Listicles can be very long – keep the amount of writing to a minimum to avoid losing the reader.
- Consider adding pictures to break up the text.

Remember to put the journalism first. A listicle can be a great way to make information seem more manageable and fun but be sure you have an appropriate reason to use the format. And consider interviewing an expert or asking other people's views to create a more compelling and credible narrative, rather than simply making up your own items on the list.

Here are some examples of stories that could come your way as a journalist, with ideas for how to turn them into a listicle format.

Table 3.4: How to convert a story into a listicle

Story	Listicle idea	Interview or additional content
A survey into rising levels of stress in young people	10 ways to combat student stress	Get tips from therapist, student counsellor, students. Link to relevant charity or helplines.
Research shows that tomatoes are good for health	Tomatoes are good for you – 9 recipes to try at home	Talk to a chef or nutritionist. Links to full recipes.
The start of the football season	7 players to look out for this season	Get a former player/manager to make their selection. Interactive messageboard to allow fans to comment on the selection.
Record levels of dogs in a local animal shelter	Meet the 11 pups looking for a new home	Interview animal shelter staff about each dog. Photographs of the dogs.
Police launch initiative to teach homeowners about safety	8 ways to keep your home safe	Talk to police officers running the initiative.
Music legend announces concert tour	6 songs by this music legend you must listen to	Get a music expert, record shop owner or fans to say why they would choose a particular track. Create and link to a Spotify playlist.

Exercise 3.8: Create a listicle based on a news story

Look at your local newspaper's website. Are there any articles there that you could write as a listicle? Use a table like the one above to think about what the listicle could be and what additional sources you could use to write it.

Try creating your own digital listicle using Word Press.

References and recommended reading

Chartbeat, 2000. *Reader traffic and engagement in 2020: By the numbers.* [online]. [Accessed 31 July 2024]. Available from: <https://blog.chartbeat.com/2020/12/16/2020-reader-engagement-data/>

Deuze, M. 2004. What is multimedia journalism? *Journalism Studies*, **5**(2), 139–152.

Hurst, N., 2016. *To clickbait or not to clickbait?: an examination of clickbait headline effects on source credibility* (Doctoral dissertation, University of Missouri – Columbia).

Hall, J. 2001. Online Journalism: A Critical Primer. London: Pluto Press.

Hargreaves, I. 2003. Journalism Truth or Dare? Oxford University Press.

Holic, N. 2014. An open letter to list articles. *Huffington Post.* [Online]. [Accessed 16 June 2022]. Available from: <http://www.huffingtonpost.com/the-university-of-central-florida-forum/an-open-letter-to-list-articles_b_4268725.html>

Kuiken, J., Schuth, A., Spitters, M. and Marx, M. 2017. Effective Headlines of Newspaper Articles in a Digital Environment, Digital Journalism, **5**(10), 1300–1314.

Medvedeva, Yulia S. 2016. Using Morphological Analysis for Final Project Planning in Multimedia Journalism Courses. *Communication Teacher.* **30**(2), pp. 67–71.

Miller, G. A. 1994. The magical number seven, plus or minus two: Some limits on our capacity for processing information. *Psychological Review.* **63**(2), pp. 81–97.

Molyneux, L., and Coddington, M. 2019. Aggregation, Clickbait and Their Effect on Perceptions of Journalistic Credibility and Quality. Journalism Practice, **14**(4), pp. 429–446.

Moran, K. 2020. *How People Read Online: New and Old Findings.* [Online]. [Accessed 29 July 2024]. Available at: <https://www.nngroup.com/articles/how-people-read-online/>

Netcraft. 2024. *July 2024 Web Server Survey.* [Online]. [Accessed 31 July 2024]. Available at: <https://www.netcraft.com/blog/july-2024-web-server-survey/>

Sadri, R. S. 2019. Listicles and the modern news article: comparing the perceived credibility of listicles and traditional articles among millennial media consumers. *Atlantic Journal of Communication.* **27**(2), pp. 83–98.

Shedden, D. 2014. *Today in media history: CompuServe and the first online newspapers.* [Online]. [Accessed 31 July 2024] Available from: <https://www.poynter.org/reporting-editing/2014/today-in-media-history-compuserve-and-the-first-online-newspapers/>

Shoemaker, P., Tankard, J., and Lasorsa, D. 2004. *How to build social science theories.* Thousand Oaks, California: Sage.

Tereszkiewicz, A. 2014. I'm not sure what that means yet, but we'll soon find out' – The discourse of newspaper live blogs. *Studia Linguistica Universitatis Iagellonicae Cracoviensis*. **131**(3), pp. 299–319.

Thorsen, E. and Jackson, D. 2018. Seven Characteristics Defining Online News Formats. *Digital Journalism*. **6**(7), pp. 847–868.

Thorsen, E. 2013. Live Blogging and Social Media Curation: Challenges and Opportunities for Journalism. *In:* Fowler-Watt, K. and Allan, S., eds. *Journalism: New Challenges*. Poole, England: CJCR: Centre for Journalism & Communication Research, Bournemouth University, pp. 123–145.

Thurman, N. J. and Newman, N. 2014. The Future of Breaking News Online? A Study of Live Blogs Through Surveys of Their Consumption, and of Readers' Attitudes and Participation. *Journalism Studies*. **15**(5), p. 655–667.

Thurman, N. J. and Walters, A. 2013. LIVE BLOGGING–DIGITAL JOURNALISM'S PIVOTAL PLATFORM?, Digital Journalism. **1**(1), pp. 82–101.

Tomkin, S. 2022. NASA unveils the first EVER images from its James Webb Space Telescope including a SELFIE – but reassures the blurry snaps are just a 'starting point' as it fine-tunes its mirrors. *Mail Online* [online] 11 February. [Accessed 30 July 2024]. Available from: <https://www.dailymail.co.uk/sciencetech/article-10502085/NASA-unveil-images-James-Webb-Space-Telescope-today.html>

Wells, M. 2011. How live blogging has transformed journalism. *The Guardian* [Online]. [Accessed 16 June 2022]. Available from: <https://www.theguardian.com/media/2011/mar/28/live-blogging-transforms-journalism>.

WorldWideWebSize. 2024. *The size of the World Wide Web*. [Online]. [Accessed 31 July 2024]. Available from: <https://www.worldwidewebsize.com/>

Digital audio

Often the focus is on video when it comes to digital journalism skills. But audio continues to be an important medium, with 72 % of news publishers identifying digital audio and podcast production as a growth area for their organisation at the start of 2023. (Newman, 2023, p5)

This chapter covers recording and editing simple audio clips, the challenge of visualising audio for use on social media, long form audio and podcasting, live audio broadcast and emerging social audio networks.

Simple audio clips for social media

Many of the principles discussed in previous chapters on digital videos and social storytelling relate equally to producing digital audio content. To start with, your story choice is important. Consider your audience, who are potentially a younger age range, and are particularly interested in stories with emotional impact, or a quirky focus.

You'll need to record in the best sound quality that you can. Your audience may be someone wearing headphones, or playing through the speaker on a laptop, phone or tablet. They may be in an office, at home, or on a train. Your audio needs to be clear and easily heard to cut through whatever might be happening in the background.

Your mobile phone contains everything you need to record an audio interview. It will have an app (such as Voice Memos on iOS, Voice Recorder on Android, Google Recorder on Pixel) installed that will allow you to easily record, save and even carry out basic edits.

Figure 4.1: Student journalist using a microphone with a windshield. [By permission of Shanai Dunglinson].

To ensure best quality you may wish to invest in additional equipment. A windshield is a cap that you can position over your smartphone's microphone. These are usually made of fur fabric or foam and can cover the whole end of the phone, or be smaller pieces that can be fixed in place using removeable stickers. A windshield breaks up gusts of air before they affect the microphone, so it will help eliminate noises like wind or breathing that might distort or disturb your recording.

An external microphone may also improve the quality of recording (see the external mic with windshield in *figure 4.1*). There are many different options and price ranges to explore. A unidirectional mic will record whatever you point it at, so it's great for voxes and interviews. Wireless will give you more range and make movement easier. Lapel mics will help you get audio of interviewees and look discreet. It's possible to get pairs of wireless microphones that will allow two people to be recorded at once.

If you're recording interviews or are concerned about using your mobile, you may wish to invest in a separate audio recorder (such as a Zoom or Tascam) that will record and store large amounts of audio.

Whatever equipment you're using there are some tips for recording

- Quiet location: Choose a spot sheltered from wind or rain. Make sure there is no distracting background noise (music in the background makes it particularly difficult to edit an interview).
- Point your microphone: Particularly when using a directional mic, point it towards the person who is speaking. That includes you when you're asking the questions, then your interviewee when they are answering. This will give you more options when editing.
- Check sound levels: It's possible to adjust the levels somewhat in edit, but if you record too loudly your sound will distort and if you record too quietly it can lead to hissing, so it's always best to check as you're recording. It's customary practice to ask someone to introduce themselves first so you can record and listen back to check everything is working correctly. Your audio should peak at around –9db.
- Wear headphones: Make sure to listen carefully as you're recording using headphones (preferably 'over the ear' style so you can concentrate fully on your recording and reduce ambient sound) so you can review your audio properly. Pay attention, not just to the subject of your interview and what they are saying, but also to the quality and levels of your recording and any background noises such as wind, music or distracting hums and squeaks.
- Background noise: Record at least a minute of 'nothing' to capture the background sound of your location. This can be handy for creating fades and fixes when editing.

Editing an audio clip for use on a website or social media is not very different from editing for a radio bulletin. Your smartphone will allow for basic editing, there are free online audio editors available (such as Audacity or Ocenaudio, for example) or any audio editing software will work. To reduce file sizes, mp3 is the format you're most likely to use to save audio for digital use.

Start with something engaging. Grabbing attention from the beginning is important. Whereas with a traditional radio interview you might start with a light warm-up question and let conversation flow naturally – or build up to a big 'reveal' – you want your social media clip to focus only on the most interesting bits.

Ensure your audio is as clear and easily heard as possible. Adjust sound levels as necessary. For this reason, avoid using background music where possible, if you must use it, make sure it's not too loud and drowning out your clip.

Keep it short. People scrolling online have short attention spans, and without the pull of video, audio posts are unlikely to keep people listening for any length of time. Different stories require different treatments, however a general rule of thumb would be to stick to less than a minute duration across all platforms – and if you can reduce that, so much the better.

> **Exercise 4.1: Practise your audio recording skills by doing an interview**
>
> Familiarise yourself with your recording equipment – whether that is your smartphone, a microphone or an audio recorder. Ask a fellow student, friend or family member to act as your interviewee. Ask them to tell you about something they feel strongly about – for example their hobby, someone they look up to or their happiest memory. Think about asking open questions and exploring their feelings to get the most honest and emotional responses. Record the interview using the tips above.
>
> When you're finished, review your audio and select a clip of around 30 seconds that will attract attention or best conveys your interviewee's emotions. Edit this clip and save as an mp3.

So now you've recorded and edited your audio clip – your next step is to decide how to present it on a digital platform. It could be inserted into a webpage (see Chapter 3 on multimedia webpages) but if you're planning to distribute via social media platforms it's likely you'll need to consider creating a visual element for the audio.

Visualisation of audio

Talking to recent graduates in industry, it's clear that social media is very much a visual medium and that the use of audio clips is seen as challenging. According to one survey of US consumers in 2019 (Verizon Media and Publicis) 69 % of people watched videos on social media with sound off and an estimated 85 % of Facebook users have the volume off. This means it can be hard to attract attention to audio content. As such it's unusual to commission and create audio content for social platforms.

Yet at times, as a journalist you may find yourself with audio that you feel is important to share. It could be a secret telephone recording that blows open a scandal, a clip from an audio interview where a politician says something revealing or a caller to a radio show with an incredible emotional story to tell.

As Jacob Tomlinson from the BBC tells us, a preference when faced with an audio story would be to go back to the guest and film again – but that's not always possible when an interview has happened in the moment.

> Ultimately, it's how to make it visual, which is a really stupid thing to say when we're talking about audio, but it's like could we have still images that we can flow over the top in relation to the story? Is there a way we've got this amazing audio that we could get back in touch with the case study and go film something with them? Can we film something in relation to it, where we can edit over? Can we do something as a bit in the past where we have word, so

> we emphasise the subtitles by using after effects and having them graphicise in fancy ways and so we're still ticking that box of being accessible for people to a hard of hearing which is great, but at the same time, it's ultimately making it far more visual.

Adding visuals to your audio content will make it stand out more on the users' feed and perform better in platform algorithms, leading to higher audience engagement.

Waveforms

An early concept to visualise audio for use on social media platforms was Audiogram Generator. Created by public service broadcaster WNYC in New York, USA, this was released as an open-source tool and has since been embraced by a number of broadcasters and podcasters.

It allows journalists to convert audio files into video, using a background image or theme and a waveform animation of the audio, which fluctuates during the video. The waveform that is created is reactive – so it accurately represents the audio in the clip, rather than being simply a moving image of a random sound wave. This creates visual interest to attract a scrolling viewer and alerts them to the fact there is audio to listen to.

Figure 4.2: Examples of using Audiogram to visualise audio clips for publication on social media.

This technology is sometimes seen as rather old-fashioned but can still be an effective way to process a sound bite for distribution on social platforms.

If you don't have the technical confidence or ability to install the original Audiogram Generator open-source software, there are several websites available to help you create the effect. Many offer a free plan, which will allow you to access most features but will include their logo as a watermark on your final creation. Search online for 'waveform generator' or 'soundwave generator' to find a full list. Examples include Kapwing, Veed, Echowave. It's also possible to add a reactive waveform using video editing software such as Adobe After Effects.

Audio cards

If you have no moving images to use, you can create a simple audio card (confusingly, sometimes this is also referred to as an 'audiogram') using a still image and audio. This is a technique commonly used by podcasters to promote their content on social media platforms.

One easy way to create an audio card for use on social media would be to use the free image software Canva. Select a template and use it to create a visual post. Choosing 'Elements' from the navigation bar you can search 'sound on' for a graphic of a speaker or waveform to highlight that there is audio. Choosing 'Upload' allows you to upload a short audio clip from your computer (this must be less than 60 seconds for an Instagram post). Drag the audio file onto your image to add it to the post. You can then download the project as an mp4 file ready for posting.

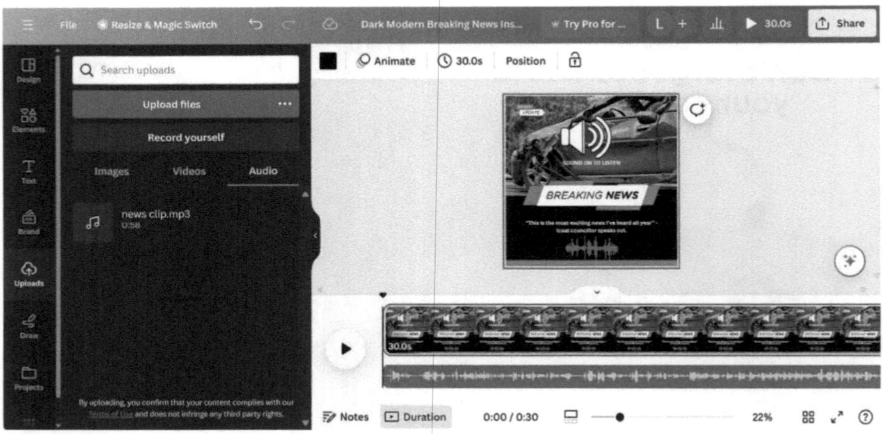

Figure 4.3: Screenshot of audio card creation using Canva.

Filming audio broadcast and interviews

The simplest way of ensuring you have the visuals you need to give your story impact on social media is to consider it from the very start. As a journalist in today's digital newsroom, it's no longer relevant to consider that your content is destined to be used on just one platform. If you're working on a story or an interview, even one that's commissioned originally for radio, think about how you can gather relevant pictures and video too.

When social media started to become a major force in distributing news, radio stations quickly realised that their audio content was not as engaging to viewers as video content. The solution – to set up cameras in the studios so that presenters and guests were filmed as well as recorded. This enabled digital producers to clip sections of the radio interview in video format, for distribution on social platforms. Many podcasters also film their recording sessions for the same reasons.

In *figure 4.4*, we can see how some radio stations have used similar but different ways of visualising their content. In this figure, LBC uses visible branding behind the presenter and for this TikTok the screen is split, Duet-style, to also show a stock image at the bottom, illustrating the news story that is being discussed. BBC Radio 1 use a close-up camera for the Instagram Reel to focus on the celebrity's face. NPR have three different contributors to their radio programme in separate locations and show all of them on camera, using a landscape format that is optimised for YouTube and Facebook. You'll note that all the examples use visible and usually branded microphones to make sure it obvious to the viewer that this is a visualisation of radio (audio) content.

Simply setting up a camera, or phone on a tripod to film your interview will mean you have easy footage to use.

Of course, this does not always work so well if a guest is 'down the line' via telephone or ISDN link, rather than on location. Originally a technology embraced during the covid-19 pandemic to comply with social distancing, the use of video conferencing tools such as Zoom and Teams by journalists carrying out distanced interviews has continued. As a journalist working on radio or print, you may consider this as your preferred option rather than a traditional phone interview. Not only is the sound quality (usually) better, it also means you automatically have visuals of the interview that can be used online. The popularity and prevalence of video conferencing during those pandemic years and since, means that many people who you wish to interview will be confident with using the technology. There are a few things to consider though, which may not be at the forefront of your mind when your first priority is getting a written quote for your article or an effective audio interview.

- An online meeting is only as good as the speed of the internet connection. Try and prevent dropouts and freezing by checking you have the best conditions. Can your guest be somewhere with better broadband coverage? If you're working from home can you ask the people you live with to hold off using

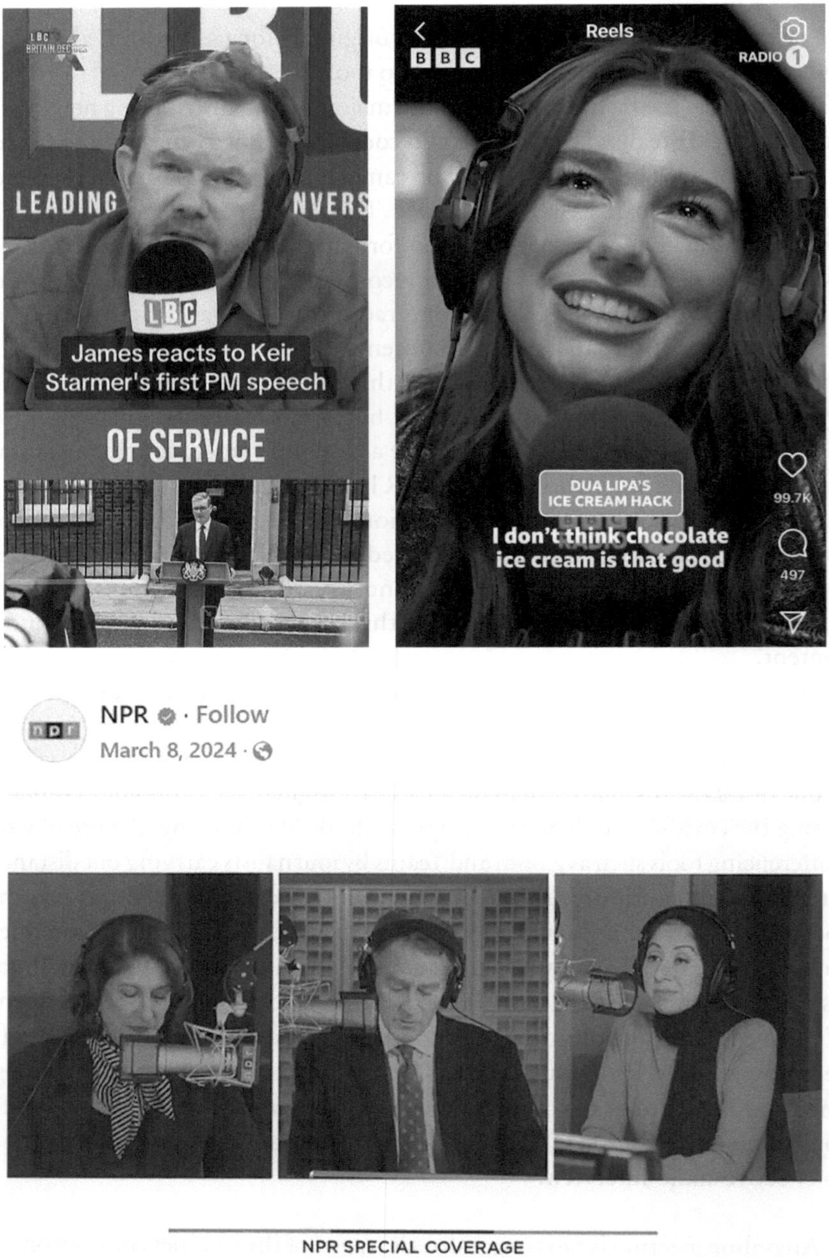

Figure 4.4: Screenshot examples of how studio cameras are used to provide visualisation of radio content.

bandwidth-hogging gaming or streaming services while you carry out the interview?

- Take some time to check the quality of your recording. Using a separate microphone can give better quality audio than relying on the computer's inbuilt one. Do you have your recording sound levels set correctly?
- It's important to take as much care over framing the shot as you would if you were filming using a camera. Check both your and your guest's background for anything that could be distracting. Underwear drying on the radiator? A poster with a swear word? Identifiable pictures of children? Any of these things could render your images unusable. Also check for any clothing issues. Your guest may not think to take the same care over dressing for a video call as for a TV interview. Ensure you're both framed in the centre of the screen, positioned so that your faces take up the same amount of screen space, in case a double shot is used. The angle will be more flattering and natural-looking if your camera is at the same level as your head – raise your computer using books if need be.

Captions and graphics

As discussed in Chapter 1, the use of captions is standard in digital video production. Not only does it help accessibility (providing a written alternative for those who may be deaf or hard of hearing), it also means the full story is available to all those people watching with their sound turned off.

If you're converting a piece of audio into digital content the use of these captions (or subtitles) is even more important. Most video editing software includes an option to automatically generate captions, transcribing the audio and synching it with your video. This includes free apps included with your smartphone, or software on your computer. There are also many online editing tools that will generate captioning for free.

If you're creating for a particular platform, the platform may have a straightforward way to add captions. For example in TikTok, after recording your video you can go to the editing page and click 'Captions' to see the auto-generated captions. For an Instagram Reel, there's a 'Captions' sticker you can use to add to your video.

It's essential that you always check captions that are produced automatically for spelling and grammar. The systems may not be able to cope with regional accents or unclear speech and could be unfamiliar with names of people and places, so there are regularly errors. Inaccuracy in your captions will reflect on audience perception of your journalism so it's worth taking the time to double check and to correct any mistakes.

Simply using large or graphic subtitles would be one effective way to visualise your audio content and draw attention to your story. These could be positioned over a relevant still or video shots. Remember that some people will be watching on a mobile device, so use colours with strong contrast and don't make the text too small or the font too elaborate.

With more resources and knowledge, more complex illustration could be used to add visuals to a story where you have excellent audio but it's pictorially challenged.

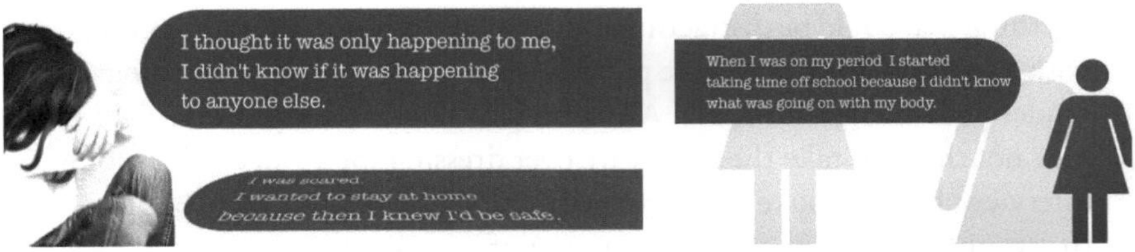

Figure 4.5: Screenshots from a digital video made for Facebook to illustrate audio interviews of anonymous school pupils discussing period poverty (BBC Radio Leeds, 2018).

'Sound on'

No matter what techniques you use to add visuals to your audio, sometimes the experience of hearing something first-hand can't be replicated. Whether it's the emotion in someone's voice, the shock of an unusual noise or a baby's first words ... the experience of hearing these cannot be reproduced using subtitles or graphics.

In this case, digital producers tend to draw attention to powerful audio by simply asking the viewer to turn their sound on for the full experience. This can be done in the written caption for the post, or as a graphic on the post itself (or both).

Of course, you can't guarantee your viewer will do as you ask, so you can't fully control the experience. This is also a technique that should be used rarely and only for the most impactful audio. You'll lose the trust of your audience and risk alienating them if you ask them to do something that they feel is not worthwhile.

Exercise 4.2: Create a visual way to illustrate audio on social media

Challenge yourself to create a visual way to present an audio clip for use on social media. You could use the clip from your previous exercise or start again with a new interview or story.

Use any of the techniques suggested above, or you may have seen something else while browsing that you want to try. Whatever technique you choose remember to include clear and accurate captions.

Podcasting

Having acknowledged the importance of video and pictures for social media, there is one digital audio format that has grown and thrived. Podcasting has its roots in audio blogs going back decades, but the first audio shows available to subscribe and download online began in the early 2000s. Since then, the growth of the format has been phenomenal.

News and journalism-led podcasts have always been popular and led the way in podcasting. Notable examples include the longest-running talk podcast Open Source (2003), creativity of This American Life (2003) and the true crime investigation phenomenon Serial (2014). Today you'll find news titles in the podcast charts in most countries around the world.

According to a study from the Pew Research Centre in 2023, almost half of Americans had listened to a podcast in the previous 12 months – and two thirds of them had heard news discussed on a podcast. This means that around a third of US adults have been exposed to at least some news via podcast.

Here we outline the basic concepts to allow you to consider a podcast as a medium for telling your news stories. There are many books looking at the production and techniques of podcasts that will offer more in-depth guidance and analysis if you're thinking of creating your own podcast.

What is a podcast?

The Oxford English Dictionary defines a podcast as a 'digital audio file of speech, music, broadcast material, etc., made available on the internet for downloading to a computer or portable media player; a series of such files, new instalments of which can be received by subscribers automatically'.

This definition suggests the main difference between a podcast and a traditional radio programme is its distribution – the fact that it's a digital file that can be downloaded and that it can be delivered automatically via subscription.

Another definition (Rime et al., 2022) introduces the fact that the listener has control of when, how and where they will hear the content. And also, the fact that it's a medium that makes content creation available to all – not just those who own or have access to a media empire: 'A podcast is a piece of episodical, downloadable or streamable, primarily spoken audio content, distributed via the internet, playable anywhere, at any time, produced by anyone who so wishes'. These combined changes in the way audio content can be delivered and consumed have led to wider-reaching changes in the way news content is conceived and created.

Podcast journalism has been described as a place for experimenting with journalistic forms, practices, and conventions (Lindgren, 2023). Podcasting is a relatively new medium and still developing and changing but already there are conventions and expectations on the format, style and delivery of podcasts that can also be very different from those in traditional media.

Here are some ways in which a podcast differs from a traditional radio programme.

- No broadcast limitations: A podcast does not have the restraints of fitting into a linear broadcast slot. This means there is no requirement for audio to be a certain length, have a particular number of episodes, etc. This allows for a great deal of flexibility.
- Democratisation of production: Anyone can produce and publish a podcast, you don't need the equipment, technical expertise, studios, budgets, etc. that are needed to produce live broadcast. This means that a podcast can be produced more quickly, more independently and in some cases more creatively.
- Audience demographics: Podcasts appeal to younger listeners more than traditional radio. Many podcasts are still developed by large production companies and broadcasters who need to meet company standards, fit company style, and make profit by attracting the biggest audience. But smaller podcasts can cater for niche markets and interests and offer more diverse perspectives.
- Tone: Listening to a podcast is often an individual experience and can feel quite intimate. Scripting and interviews are often informal and using an 'authentic' voice offering the listener a chance to feel included and welcomed.
- Reporting style and storytelling: A podcast often features personalised journalism with self-reflexivity and transparency of the reporting process (Lingren, 2023). This is a very different experience from the journalistic norm of the objective reporter, emotionally distanced from the story and reporting only the hard facts. Podcasts also often offer the listener the chance to follow the journey of how a story is investigated or glimpse a behind-the-scenes insight into the way news is made.

The lines can be blurred, with many successful podcasts (such as *This American Life* from WBEZ Chicago) starting life as a radio programme offered in a downloadable form as well as on linear transmission. Equally the success of podcasts has led to many traditional radio programmes reflecting some of the same stylistic decisions and storytelling techniques (intimate style, informal tone, transparency of journalistic process, etc.) that have been popularised by podcasts.

News podcasts

News podcasts are popular with listeners, with news, politics and current affairs podcasts regularly appearing in global download charts. It's useful to refer to a typography developed Newan and Gallo (2019) which identifies five general types of news podcast.

News roundups: Short podcasts designed to give the listener an update on the day's news and current affairs, usually produced and published daily. Examples: *5 Things* (CNN), *Up First* (NPR), *World in 10* (The Times), *The Smart 7* (Daft Doris)

Deep dives: Longer podcasts (20 minutes) using audio documentary techniques, narrative, clips and sound to explain one or two stories in more detail. *The Daily* (New York Times), *Today in Focus* (The Guardian), *Sky News Daily* (Sky News), *Consider This* (NPR), *The Monocle Daily* (Monocle Radio)

Documentary: Narrative series, with multiple episodes (30–40 minutes) investigating a single subject. Examples – *Serial* (Serial Productions), *The Missing Cryptoqueen* (BBC), *The Trial of Lucy Letby* (Daily Mail), *The Teacher's Pet* (The Australian), *Sweet Bobby* (Tortoise)

Talk/interview: Longer episodes of personality–led informal, round table extended discussion on news topics. Examples – *Newscast* (BBC), *News Agents* (Daily News), *The Tucker Carlson Show* (Tucker Carlson Network), *The Joe Rogan Experience* (Spotify)

Audio read: Reading of an article published in a newspaper or magazine. Examples – *The Long Read* (The Guardian), *NYT Audio* (New York Times)

Of course, there are other formats and ways of expressing news in a podcast – and new ones yet to be discovered – but these broad groups cover the majority of news podcasts available to listeners today.

Exercise 4.3: Listen to and analyse podcasts to identify winning formulae

Listen to news podcasts in one or more of the categories above and give a critical analysis. What works? What don't you enjoy? Who is presenting and what personality do they bring to the discussion? How is the podcast structured? What audio techniques do they use? How do they keep you interested?

Storytelling

Some podcasts have such an informal and chatty presentation style, it feels as though they are recorded off-the-cuff in one go – however, it's extremely unlikely that this is the case. A successful podcast is carefully planned, targeted and edited. Even the most relaxed, friendly or seemingly chaotic of podcasts should have a sense of purpose, an understanding of the target audience and a structure that engages and tells a story. The more time you put into planning your podcast, the easier it will be to produce each episode.

Here are ten things to think about when planning your podcast

1. Purpose: What do you want your podcast to achieve? Do you want to showcase material from elsewhere, reach new audiences, tell stories that are not easily accessible, demystify a complicated topic, create a sense of community?
2. Audience: Who will listen to you? A podcast could be reached by anyone around the world, but the more specific you are about who you are aiming at, the more you can refine your content and style to attract that listener. In general podcast listeners attract a young, well-educated and better off audience (Digital News Report, Reuters, 2024).
3. Format: Does your podcast fit into one of the categories outlined above? A podcast is episodic – with listeners expecting the same format each time they listen. How will your podcast be structured and how will it develop over the series of episodes?
4. Familiarity: You want your listener to subscribe to your podcast and return for episode after episode. How can you give them a consistent product where they know what to expect? You could build familiarity by having a strict format, or by introducing a regular feature such as posing the same question in each episode.
5. Voices: Who will present your podcast? If you have multiple presenters, consider what it is that each brings to the table. Perhaps expertise in a particular area or a different viewpoint. Think about the guests or contributors you need to make your idea work.
6. Storytelling: Think about the narrative you're trying to tell – whether it's a different one for each episode, or a long story stretching over multiple episodes. Where will you start, where will you end? How can you plan the story in a way that keeps the audience learning more and interested? What are the key points you want to get across in each segment or episode? Make sure you're taking the listener with you on a journey and not confusing them or boring them with a lack of purpose.
7. Standing out: What makes your podcast different from everything else available? Original topic, unusual format, incredible guests, well-known presenters? Make sure that this is clear and highlighted in the design of your podcast. It's estimated that there are over 4.2 million podcasts, with thousands of shows published each day across the world (podcastindex.com, July 2024). You need to clearly understand and communicate to your potential listeners what makes yours worth listening to.
8. Show summary: You'll need a brief description of your podcast to use when you publish it. This information is how potential listeners will decide

whether to click on your podcast or not. It needs to be concise but get across some crucial factors such as who the show is for, what they will get out of it and why they should listen. Use keywords that might attract your target audience's attention. Think of this summary as an 'elevator pitch' – a few short sentences that would 'sell' your show to someone if you only had the time it takes to travel down a few floors in a lift.

9. Podcast name: The name of your podcast to be short so it fits onto a screen and is easily seen in listings. Most podcast titles are fewer than five words long. Don't be too obtuse – choose a title that describes the podcast so potential listeners know what to expect. Think of your target audience and words that may attract them. Check that there are no current podcasts with the same or a similar name.
10. Thumbnail image: The logo or visual identity is another way to attract the audience and make your podcast stand out. Spend some time thinking of a simple image that represents your podcast and will work when shown in a list of images on a small screen. Many podcast logos include the name of the podcast in text – make sure this is in a size and colour that make it legible.

Exercise 4.4: Create a podcast pitch, explaining what will make your podcast work

Planning is key to an effective podcast. Take time to think about what your podcast will sound like, who will listen and how your podcast idea would work over a period of time, to encourage people to subscribe. To do this, work on a podcast pitch.

You'll want to include:

- The name of your podcast
- Your podcast summary – try and write this in one or two sentences
- Decide your target audience
- Details of presenters/hosts
- A template running order that each episode will follow
- Brief examples of theme/content/guests for up to four episodes
- A rough idea for the thumbnail artwork that will be used to promote the podcast

If you're working in a group, you can then present your pitches to each other and decide which podcast you would be most likely to commission.

Production

It's important to remember that your podcast is subject to the same legal, regulatory, and ethical considerations as any other piece of journalism that you produce. In addition, as podcasts are distributed online, they are available to listeners across the globe. This means you could face legal issues in any country. Make sure that you have fully researched and checked your story. If your podcast is not fully scripted make sure you listen to everything before publishing to check for accuracy and any potential legal concerns.

Before recording you'll want to produce a running order for your podcast episode. Even if you're making a conversational podcast, it's important that everyone involved is clear about the story that you want to tell, and the points you need to hit to get there. Be clear about this before starting so that everyone knows what is expected. It may help to map it out on paper first for reference.

Table 4.1: Example running order for a podcast

Time	Segment	Notes	Script
00:00	Intro	Audio: intro.mp3, tease.mp3	(INTRO MUSIC – intro.mp3) Pres 1: Welcome to the latest edition of our Animal News podcast Pres 2: Each week delve deeper into the news stories that have caught our eye featuring our friends with fur, feathers and fins … Pres 1: And this week we're meeting a woman with an amazing tale to tell about an encounter with an octopus (CLIP – tease.mp3)
01:30	Headlines	Guest in studio	Pres 1: But first we're joined by our resident vet to take us through some of the stories she's spotted in the headlines this week. Pres 2: Did you notice the story about the runaway kangaroo?
07:00	Octopus	Audio: love_music.mp3, octopus_int1.mp3, bubbles.mp3, octopus_int2.mp3	
10:00	Story 3		
15:00	Wrap up		
18:00	Outro		That's it for this week's podcast – next week we'll be back with more

To script or not to script?

How fully you script in advance will depend on the type of podcast you're making. A brief podcast may need to be fully scripted to keep to time. A documentary may need to be clearly scripted to ensure it's legally compliant and tells the story fully in a clear and entertaining way. However, for a conversational podcast it will sound more natural if your presenters do not read out lines from a script. Your news content needs to be as well-researched and as accurate, whatever the format you choose. If you're not scripting you still need to be clear about what you're covering and make a note of important facts you might want to refer to, quotations you might make that need to be accurate and of course, any legal concerns you need to avoid. Having a running order and a list of bullet points to refer to will help organise your natural conversation.

It does make sense to carefully script the start and end of your episodes. You do not know where the listener will start their journey with your podcast. Sometimes they will search out episode one and start at the beginning, but often their first time to hear you will be the most recent episode, your most shared episode, or one that happens to cover a topic they are interested in. So, you have to lay out your stall, introduce the podcast's purpose, the presenters, any key facts every time. Often this means the same introductory script is repeated at the start of each episode, perhaps with small tweaks to explain the current episode. Being able to clearly explain your podcast's purpose in just a few informal sentences is vital. You may find your podcast summary (see above in 'Planning your podcast') to be helpful as a starting point. Podcasts work on a subscription model, you want your listener to keep listening and to sign up to future episodes. The end of your podcast is the place to tease towards the next one.

Gathering audio

Before you record your podcast, prepare all the audio you need in advance.

If you've carried out any interviews or prepared other content in advance, check and edit it first so that the final recording process is as simple as possible.

Think about any natural sound effects you may need to gather to help to tell your story. These can be effective to build a soundscape and take the listener on a journey with you.

You'll probably want to select some music to act as an intro or outro for your podcast. Choose something short and catchy to use as your signature theme. You may want to use music elsewhere to build tension or another emotion.

Remember that any music or audio you use will be subject to copyright laws. Ask for permission, or check copyright and fair use policy, before using any audio from a third-party source in your podcast. Perhaps you or a talented friend are able to compose original music? Otherwise, there are many websites offering copyright and royalty-free audio that you could download for your projects (try Uppbeat or Pixabay for example). Remember to read the terms and conditions of use carefully.

Recording

A podcast could be recorded anywhere from in your home using just your phone to using professional equipment in a specially designed studio. We've already discussed simple tips for recording and editing audio at the start of this chapter. As a general rule ensure that all your audio is recorded in as good quality as possible. If you have multiple presenters or interviewees then recording everyone in person together (rather than having online contributors) will make for a more natural conversation, and using a separate microphone for each person will help with quality.

Location is important – if you don't have access to a recording studio, ensure a quiet room with as little background as possible. A podcast can be a lengthy piece of audio and distracting noises or inferior quality audio will be off-putting for the listener. Listen out for and eliminate any background noise including hums from electronic equipment such as fridges, computers or air conditioning. Drawing curtains and even hanging blankets against walls could help dampen any echoes.

Figure 4.6: Example of a specially designed studio, University of Leeds.

Consider the tone of your podcast before recording and discuss with all the presenters so everyone is aware. Do you want your podcast to feel like a chat among friends, an informative discussion panel, a private conversation?

Many podcasts use an informal and personal tone. This may mean moving closer to the microphone than you would do for other forms of presenting. Doing this will allow you to speak more quietly and conversationally, rather than projecting your voice, creating a more intimate experience for the listener.

It may also help to make physical adjustments to your recording space and how you sit to create an environment that suits the tone. For example, sitting in a more relaxed position, dimming the lighting may help create a relaxed atmosphere if that's the tone you're hoping for. On the other hand, sitting upright, or even standing, could produce a more energetic style.

Editing

Be ruthless with your editing. While a podcast is not bound by any set length, that doesn't mean you should go on and on! If you lose your listener's interest it's easy for them to tune out, turn off, and not come back for another episode. So, you need to keep them entertained. Only keep the very best bits that you need to tell the story. Cut out any waffle or conversation that doesn't add anything. Although you'd hope your listener will stay with you for a whole episode, that isn't always going to be the case. So don't be tempted to build up to a big reveal and leave the best bits to the end, make sure you're at least teasing your highlights from the very start of the episode. Using multitrack editing software you can add in clips, music or sound effects as needed.

Common practice would be to record and save your audio in a high quality format such as .wav, but then save your file as an .mp3 for distribution. This is a compressed format so file sizes are smaller. You'll lose some audio quality during the compression process, but this is considered a reasonable sacrifice to ensure the podcast is quicker for people to download and takes up less space on their phone or computer.

Dissemination

Once you've recorded your podcast you need a way to get it out to listeners. Firstly, you need somewhere to store the files online for access. Then you need to create an RSS (Really Simple Syndication) feed. This is a piece of code that stores details about your podcast, such as where the file is, what it's called. This RSS feed can then be added to podcast directories (such as Apple, Spotify and Google) to list the most up to date information about your podcast. This is how people will find your content.

There are many podcast hosting services available where you can upload and store your podcast and then automatically generate the RSS feed and distribute details to popular directories. Some of these (examples include Podbean, Buzzsprout and Spreaker at time of writing) offer the service for free. Check to see what exactly a podcast host provides and what you need. There may be limited storage, bandwidth, analytics and monetisation opportunities with free services. It's always possible to swap hosting if your needs change.

As previously mentioned in this chapter, it's also increasingly common to see podcasts filmed as well as recorded – allowing distribution on additional channels (YouTube) and meaning video clips of 'best bits' can be taken for social media as a promotional tool and to reach new audiences with content.

Social audio

Audio has been a part of social media for a long time, but the meteoric rise of Clubhouse in 2020 saw a huge spike in interest in audio chatrooms. Experts have put the popularity of Clubhouse and similar platforms down to the covid-19 pandemic and the desire for people living under lockdown restrictions to make connections with conversations more personal than written texts, but less intrusive than video as 'Zoom-fatigue' set in (Zhu, 2021).

It certainly seems that popularity has waned since the peak of the global pandemic. After a huge surge to 10 million in early 2021, the number of Clubhouse users had dropped to 3.5 million by September of that year. Many other short-lived social audio apps have since launched and have not taken off. However there does still seem to be increased use of voice features in social network platforms such as WhatsApp and Discord and new audio-only social platforms continue to be launched (e.g. Airchat, 2024).

Major social platforms are also starting to experiment with social audio. Facebook, Slack and Spotify have all developed audio channels for live discussion. X has an area called 'Spaces', launched in 2020, where users can start a live public audio stream. Anyone can join and listen (with the option of doing so anonymously) and up to thirteen people can speak at any one time, with a host and co-hosts managing how this works. It's a feature currently not particularly used by major news corporations to host conversations, however journalists listening into Spaces can find and report on news stories. For example in July 2024 Kenya's President William Ruto hosted an X Spaces conversation to engage with young people following anti-tax protests which began on social media. A reported 150,000 people listened in (BBC, 2024).

As a relatively new concept, there have been concerns over privacy, safety and misinformation on audio social platforms. As yet major news broadcasters are keeping an eye on, rather than investing in, content for them. So, the jury is still out on the future impact of audio social media platforms. A challenge for journalists in the future may be to watch how these continue to develop and consider what opportunities they may provide for sourcing and publishing news stories.

Exercise 4.5: Experiment creating content for a social audio platform

Familiarise yourself with a social audio platform and consider how you could use it to find or tell news stories.

To find Spaces to listen to, look in the main menu of X. You can join as an anonymous listener, or share your details and participate in the conversation. You can even start your own Space conversation by holding the plus button where you would usually do a short press to add a post. Click on the Spaces microphone icon. It's possible to start immediately with a live stream, or schedule when you want a conversation to begin so you can invite people to join you at a particular time.

References and recommended reading

Fleming, L. and Muia, W. 2024. Kenyan leader faces furious young people in online debate. *BBC News*. [Online]. 5 July 2024. [Accessed 7 August 2024]. Available from: <https://www.bbc.co.uk/news/articles/ce58y8ngdk5o>

Lingren, M. 2023. Intimacy and Emotions in Podcast Journalism: A Study of Award-Winning Australian and British Podcasts, *Journalism Practice*, 17(4), pp. 704–719.

Newman, N. 2023. News podcasts: who is listening and what formats are working?, *Digital News Report 2023*, Reuters Institute for the Study of Journalism.

Newman N. 2023. *Journalism, media, and technology trends and predictions* 2023, Reuters Institute for the Study of Journalism. [Online]. [Accessed 11 August 2024]. Available from: <https://reutersinstitute.politics.ox.ac.uk/journalism-media-and-technology-trends-and-predictions-2023>

Newman, N. and Gallo, N. 2019. News Podcasts and the Opportunities for Publishers, *Reuters Institute for the Study of Journalism*.

Rime, J., Pike, C., Collins, T. 2022. What is a podcast? Considering innovations in podcasting through the six-tensions framework. *Convergence*. 28(5), pp. 1260–1282.

Shearer, E., Liedke, J., Matsa, K. E., Lipka, M. and Jurkowitz, M. 2023. *Podcasts as a source of news and information*. Pew Research Center.

Weldon, G. 2021. NPR's *Podcast Start Up Guide*. New York: Ten Speed Press.
Whipple, K., Ashe, I., Cueva, C. and Lourdes, M. 2023. Examining Podcast Listeners' Perceptions of the Journalistic Functions of Podcasts, *Electronic News: Broadcast and Mobile Journalism*. 17(1), pp. 40–66.
Zhu, B. 2021. Clubhouse: A popular audio social application, *Proceedings of the 2021 International Conference on Public Relations and Social Sciences (ICPRSS 2021)*. Atlantis Press.

Understanding audience data

This is a new twist on an old challenge. It's always been necessary for journalists and news organisations to try and understand their audience; to serve current readers and viewers as best they can, while looking for opportunities to increase their share of the market. With legacy media, it was relatively straightforward for news companies to know how the audience was responding to their efforts because they could track sales (for print) or viewing figures (for radio and TV). That's because with 'old media', the news producers controlled both their content and the means of sharing it, with newspapers publishing and distributing their hard copies, and radio and TV news services being broadcast by their owner's station or channel. This power was partly retained in online publishing through proprietary news websites, though publishers had no control over how and why pages were ranked by search engines.

But it's in the growing area of social media journalism that the picture changes most dramatically. CNN, the BBC and the Sydney Morning Herald, along with all other news outlets around the globe, are at the mercy of third-party tech giants when it comes to distributing the kind of native content that you have learned to produce in previous chapters. The social media sites act as publishers, dictating what is seen by who, where and when. They also control the data that goes with the consumption of social stories, that give vital insights into who is interested in what. To get to your social audience, you need to go through the platforms.

The challenges this creates were on the minds of the recent Journalism graduates we spoke to who are now working as digital journalists. They confirmed that understanding social media and its users are bound together. They talked about the importance of being able to get into the mindset of the audience, through getting a handle on the platforms and the differences between them.

All of which is crucial to your approach to production because, as Georgia from Channel 4 exemplified it, 'you'll tell the same story in completely different ways for Facebook and TikTok'. In this chapter, you'll learn key differences between social platforms, how to use audience metrics to measure the impact of social and online journalism, and the role algorithms play in determining how many people get to view our work. All of which has consequences for the role of journalism in society, news values and editorial standards, that we encourage you to explore at the end of the chapter.

Social media users

Demographics

Let's start by looking at key social media demographics to get an idea of how platforms differ and who uses them, with a few caveats in advance. As we're now familiar with in the world of digital journalism, things don't stand still and that's true when it comes to social media use; what we provide below is a recent snapshot. Social platforms are all relatively new and so demographics will change and we've seen that already. There was a time when Facebook was for young people and as those young people have become older people, many have stuck with this platform and not taken up new social media. There's a form of generational loyalty at work and we're likely to see that with social sites that are currently preferred by today's young folks. It's also an over-simplification to say that there's just one audience on each network, so what *table 5.1* suggests is where the audience is strongest. Our focus is on the five social media that have established or, in the case of TikTok, fast-growing use as a news destination. Finally, we've included figures specifically on teenage use of social media as this group is often an indicator of future trends.

These basic evaluations only provide a high-level overview of a small set of considerations, but they give you an idea of who goes where on social media and the sort of considerations you need to make. Throughout this book, and notably in the social short section, you've learned about how story choices are impacted by demographics. They can be cross-referenced with the character of the social media to help you decide how to tell your story. As an example, you might derive from the top row of *table 5.1* that, with a growing news audience and its popularity amongst young adults and teens, you could target Instagram for a report on education. As it's an image-based platform with a growing emphasis on video and photo carousels, think about how you can tell that story in a visually impactful way.

Table 5.1: Characteristics of a selection of social media and their audiences

Social media	Character	News use* (arrows show change from previous year)	Demographics **
Instagram	Photo and video sharing	15 % ↑	Age: 18–29 year-olds Teenage reach: 62 % Gender reach: Women 54 %, men 39 % Income: Middle-high earners
TikTok	Video sharing (mainly short form) with audio linking	8 % ↑	Age: 18–29 (62 % with big drop to 39 % at 30–49) Teenage reach: 67 % Gender reach: Women 40 %, men 25 % Income: Little variation
X (formerly Twitter)	Professional and political social networking	11 % →	Age: 18–29 Teenage reach: 23 % Gender reach: Women 19 %, men 26 % Income: Very high earners
Facebook	Personal social networking	26 % ↓	Age: 30–49 Teenage reach: 32 % Gender reach: Women 76 %, men 59 % Income: Middle to high earners
YouTube	Short and long form video sharing	22 % ↑	Age: 18–49 Teenage reach: 95 % Gender reach: Women 83 %, men 82 % Income: Increasingly popular from low to high earners

**According to the 2024 Digital News Report by the Reuters Institute, based on an international audience sample asked to name social media they have you used for news in the last week.*
***Age and income figures indicate where each social media performs best, sourced from 2024 Pew Research Center findings based on a sample of adults in the USA. The gender reach figures are from the same research and the teenage reach figures are from another Pew report, based on a sample of 13–17 year-olds in the USA.*
References to these sources can be found in the references list at the end of the chapter.

Our interviewees made it clear that, as is typical in the industry, the large media organisations that employ them invest a lot in finding out what works on different platforms, particularly in relation to their own output. This tells us that, as an individual reporter, you are unlikely to be asked to carry out market research or be the person who derives a social media strategy from the findings. So, what we promote here is that you have a good understanding of the principal differences between social media and the processes news organisations go through to keep on top of what's required to ensure that your journalism is seen.

Editorial impact

Two University of Leeds alumni, who now work for Channel 4 in the city, explained how the differences between social media impacts on editorial decision-making and production. Here are four examples of how it manifests:

1. Story selection: Georgia points to editorial decisions being affected by the audience. For example, the young adult to middle-aged prominence and its popularity amongst women, has created a 'stereotype of the 'Facebook mums' This contrasts markedly with the much younger audience on TikTok. She says that this 'massively dictates the content you're going to pick to put on the platform. Facebook favours emotional, human stories, and the videos themselves can be fairly simplistic. TikTok is much more humour based', meaning quirky and light-hearted stories are a good fit.
2. Production decisions: The audience, alongside the character of the platform, affects how reports are created too. Georgia says that TikTok 'videos need to capture attention with more blatant methods, like using moving graphics, large, colourful subtitles and effects, to make sure there is enough going on on the screen to keep the viewer from instantly scrolling on'.
3. Sound: Georgia also highlights the contrasting significance of sound between TikTok and Facebook. The former is largely a music-based platform, 'where videos go viral for using specific sounds and hopping on trending audios'. By contrast, most viewers on Facebook consume videos with the audio muted. This has implications for captioning, which is clearly vital for Facebook users: 'It also means that on TikTok, you might want to lead with the best sound bites, in the hopes that the sound might be repurposed by other users'.
4. Duration: The length of videos provides headaches for digital journalists as what the platforms want, i.e. what their algorithms promote, is inconsistent. However, Jack, who works in the sports division of Channel 4, identifies some of the main differences between social media. He explains that YouTube functions somewhat like a TV service, with videos of 8–14 minutes performing well on their channels. The same piece would be cut down to something in the region of four minutes for the adult audience on Facebook. At the other extreme, Instagram Reels and TikToks are kept to impactful moments, that could be as short as 10–20 seconds.

The following exercise helps you get your own feel for how news organisations put into practice what they understand of social media audiences. It will give you a snapshot of how things are now and you can return to your findings in future to help you spot any new trends.

Exercise 5.1: Compare how the same stories are told differently across social media

Compare how the same story is told by the same news service across their digital accounts. Choose a big current topic that is sure to have been extensively covered on social media and online, then make observations and analyse how the storytelling changes:

Think about how the story is promoted via the cover, thumbnail, headline or initial post.
The narrative structure is important, particularly what's at the start because this is the bit that the journalists believe will hook the audience. What do you notice?
Consider the media that they use and how it relates to the character of the social media.
Production techniques, such as the pace of editing, use of audio, inclusion of captions, are also significant in relation to who they believe is their audience.
Does the writing style change and, if so, why?
Think also about contributors. Who gets to have a voice in the report and how prominent are they?

Make a list of your observations and analyse them to make comparisons. What's different, what's the same and how do the editorial decisions relate to the chosen platform and audience?

We end this section with more good news from Georgia, to remind you that what you bring to the table gives you an edge in the eyes of employers:

> Being a digital native, understanding and using the platforms is probably more important than the technical side of things because that can be learned on the job. Understanding the platforms is fundamental and I think that comes from the years of having used them'.

Over to you ...

Analytics

We've seen how audience research is central to knowing the social media platforms and finding out who's using them. What we're going to look at in this subchapter is data that indicates how well the content we create is performing. It's important because metrics 'are the quickest and most concrete way to show how many people you and your work are reaching. They are solid proof that you have an audience' (Barron, 2022). Publishers and social media companies harvest huge quantities of data in real time about who is viewing their work. This raw usage data is processed using analytical tools to provide news organisations with a vast amount of information about their viewers – where and who they are, what and when they view content and for how long, what they are interested in and how they choose to interact with it. These insights are interpreted and fed into editorial and publishing strategies.

Analytics have been touched on elsewhere in the book in relation to editorial judgements in digital journalism. In this section, you'll get a closer look at what metrics are gathered and how they are used. Whichever media company you work for in future will have its own bespoke set of data gathering and analytical tools that you'll learn about on the job. We'll focus here on some of the key measures common to all digital analytics, show you what audience data looks like and consider how it's interpreted in newsrooms to form judgements about which stories are told and how.

Website analytics

We'll start by looking at a set of analytics for a news website. The image in *figure 5.1* is an example dashboard from a supplier of online analytical tools called Chartbeat (Google Analytics is similar and available with a Gmail account for a subscription fee). It highlights, for a small news website in this instance, their *top stories*, where their users live (*location*), what *devices* they're using, the *frequency* of visits, how long users stay before clicking on a new page (*engaged time*) and how people are finding the site, known as *traffic* (i.e. whether they enter via a search engine, social media or going direct to the website). Another significant metric, not represented in our example dashboard, relates to the *timing of posts*. Analysis shows that there will be times of the day on different days of the week when a website receives a higher frequency of visits and more engagement. Knowing when these peak times are allows news organisations to plan when they publish stories to reach the biggest audience possible.

Another example of how findings from the analytics might affect editorial strategy, is looking at which specific sites are sending readers to your webpages. In the list of 'Referrers' in the Chartbeat example, Facebook is pushing almost double the amount of traffic to the news site than the next best social media. Aware of this success, the publishers might try to build on it by prioritising posting to Facebook over other platforms. In the 'Device' section, we see that the majority of viewers are connecting to the site from their mobile phones, indicating the suitability of a page layout that suits the small screen and portrait orientation of a smartphone.

Social media analytics

The same analytical tools can be used to measure the performance of your social channels. The social media companies themselves also offer access to data within

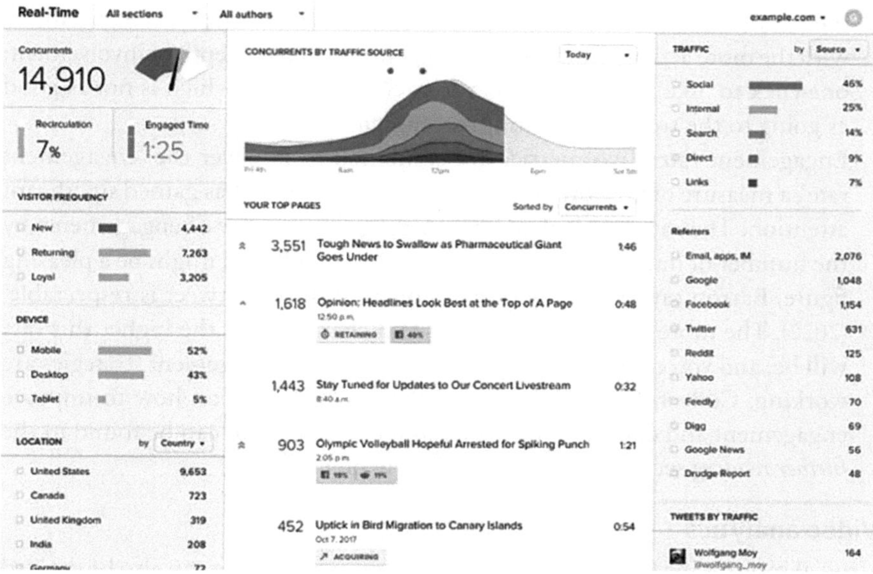

Figure 5.1: Screenshot of a Chartbeat dashboard displaying analytics for a small, local news website in the USA.
Source: Software Advice UK.

their apps, although you won't be able to get to it with a normal personal account. On Facebook and Instagram, the 'Insights' section is available if you have a business page and the 'Analytics' provided by X require you to pay for a premium account.

Wherever publishers get the information from, some of the metrics important to the analysis of social audiences differs from what we've seen for websites:

- Impressions and Reach: Both metrics capture how widely your content has been seen, though they're subtly different; impressions refers to the number of views in social feeds, whereas reach measures the number of unique users who saw your content. Although they give an indication of the success of your work, they don't provide an insight into how viewers valued your posts and how much they took in.
- Engagement: Similar to 'engaged time' for websites, this metric is used to find out how many users not only saw your content but took a genuine interest in it. It's a count of how many people 'engaged' with your work by clicking on the screen to perform an action, such as sharing, reposting, liking, commenting, mentioning or reacting with an emoji. Analytical tools can provide a visualisation of overall engagement and/or a breakdown of each of

these actions. That's helpful because there's a kind of hierarchy of engagement, with the more activity from the user suggesting a greater depth of involvement; one-click to 'like' is not as good as four clicks to share, which is not as good as going to the trouble of writing a comment.

- Engagement rate: Two metrics are combined to discover the 'engagement rate', a measure of how often your social media content has gained significant attention. The rate is calculated by dividing the number of engagements by the number of impressions. To give an indication of what might be a pleasing figure, Barron says 'anything over 1.4 on an individual tweet is respectable' (2022). The more engaging your posts and channels are, the higher this rate will be, and you can track it over time to see if your engagement strategies are working. Colleen Christison has written for Hootsuite on how to improve engagement and details of her article and free calculator can be found in the *further reading* section at the end of this chapter.

Video analytics

Crucial insights into how people view videos on social media can also be derived from studying metrics. The number of views a video receives is only part of the story because the analytics have shown that people rarely watch them until the end (rude!). Have a look at the contrasting *audience retention data* for two news videos posted to the same account.

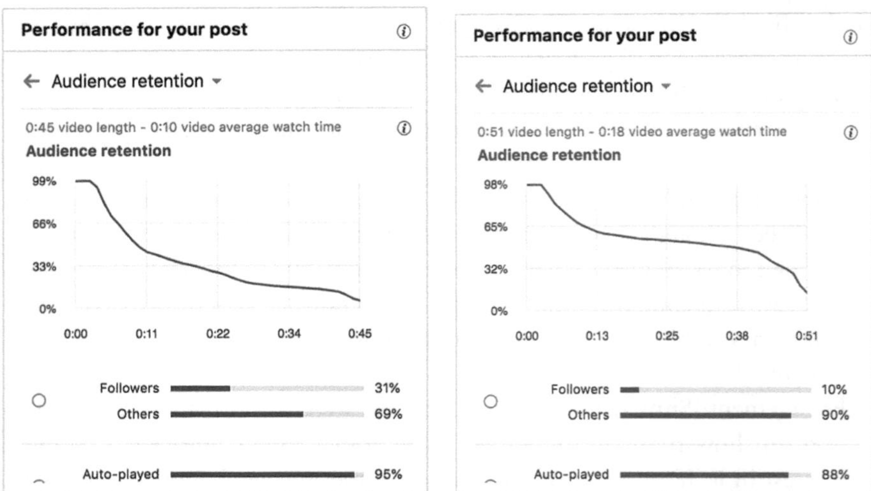

Figure 5.2: Facebook Insights audience retention data for two different videos posted to the site by the same news producer.

The video represented on the left lost nearly a third of its viewers in the first five seconds and by around ten seconds, less than half the original viewers were still watching. This is typical for videos on social media, as we scroll from one post to the next, looking for something that we're truly interested in and willing to give a minute of our lives to. The audience retention for the video on the right is much better, with around two thirds still watching at the ten second mark. The difference in performance is also highlighted in the average watch time of 10 and 18 seconds respectively.

This type of analysis provides the newsroom with food for thought: 'If the analytics tell you the audience is dropping off at a certain time, try to figure out what's turning them away and adjust accordingly' (Bean, 2021). In our example, the publishers put the success of the video on the right down to the jeopardy inherent in the short film, as it depicted a plane landing uncertainly in heavy winds. It touches down safely at around 40 seconds and that's where we see a big dip in the graph.

You won't always be dealing with the social video gift of a plane in peril, but how about teasing the audience at the beginning with juicy content that you hold back for later in the video? Whether it's possible to do that or not, this data underlines how important it is to get off to a strong start and grab the audience's attention immediately. It also reenforces the need to maintain the intensity of the narrative to keep people watching for longer, an approach advocated in the digital video chapter by our industry specialists.

One place you can get access to video analytics without needing a business account or subscription, is YouTube. You can find an analytics page in the 'Your videos' section if you have a YouTube account. It covers metrics such as *views* for each video and your channel, *average viewing times* and what percentage of the full video they represent, how people found your videos (i.e. *traffic*), *impressions*, *reach* and, if you receive sufficient views, *demographics* too. This is a great way to get to know your audience and a place where you can experiment with your video strategy to see what impact it has on the numbers. It's worth noting that audience retention on conventional YouTube videos (i.e. not Shorts) is greater because people often go to the platform with the intention of watching specific videos, rather than coming across them amidst other media in their feeds.

You can also see some of the metrics for professional news outlets without paying for industry tools. There are examples in *figures 5.4 and 5.5*, the first showing a BBC Yorkshire Facebook post that accompanied a highly popular video about the funeral of a local sporting hero. The video was played two and half million times and engagement is also high, with around 19 thousand reactions and a thousand comments.

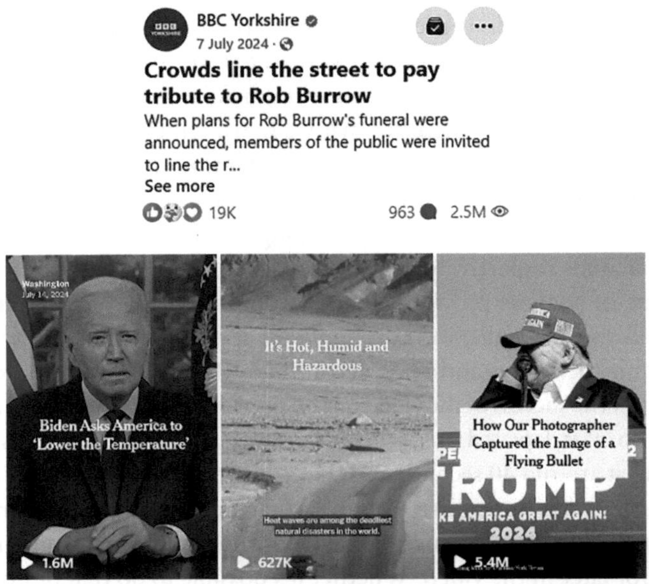

Figure 5.3: Publicly available metrics at the bottom of social media posts.

The image panel alongside it is a screengrab of New York Times Instagram Reels from July 2024, days after an assassination attempt on presidential candidate Donald Trump. The numbers at the bottom of each cover tell us how many times each video was played and the one on the right, with text promising quality photos of the attack, has far the highest views. This could be an instance of the storytelling form being aligned with the character of the platform, as discussed in the previous subchapter. People go to Instagram for imagery, so it shouldn't surprise us that an image-based report on a huge news story has done well.

Exercise 5.2: Use publicly available metrics to analyse news on social media

Investigate why some stories receive more views and engagement than others by looking at the performance of posts to social media by professional news organisations. As we've illustrated, you might look at the number of plays for Reels on Instagram or the amount of engagement posts receive on Facebook. Focus on publishers that interest you and look to the extremes for evidence of work that has either been very well received or relatively ignored. Consider what the reasons might be behind those contrasting fortunes and how the organisation might adjust its social storytelling strategy as a consequence.

A final suggestion is that you keep an eye on the performance of your own social media reporting as part of the process of learning how to interpret analytics and include them in your reflections on the work that you have created. As with every other aspect of your learning, you should reflect on it afterwards, make observations about what you experienced, think about what went well and what didn't and why that might be the case, and analyse those findings. The basic metrics available to you on your own social media accounts should be part of this process.

Understanding algorithms

But all this work on audience metrics is complicated by ever-changing third-party algorithms that dictate who gets to see our stories. Algorithms in digital media apply rules by which content is ranked, according to criteria prioritised by the platforms. The higher the ranking your content has, the more likely it is to appear in web search results or social media feeds. How these algorithms are configured, therefore, has a huge effect on metrics, our analysis of them and the conclusions we reach about the merits of our reporting.

The big tech companies do not usually share what their algorithms are prioritising with news organisations or other businesses, which causes frustration in the newsroom. Jacob from the BBC's social team, says, 'it's about working out what content has done well for you and the themes and correlations between that content'. He uses 'video audience retention' as an example of a metric that they might discover, through trial and error, has been heavily weighted by the algorithm, i.e. videos that keep people watching for a long time are given prominence in users' feeds. When that's the case, it switches the team's focus to getting 'people watching for as long as possible because we see reward on these social platforms'. But Harcup underlines how this experimentation is 'a bewildering process, partly because it can change by the second and partly because it is not always clear which aspects of particular stories have led to their apparent popularity' (2015, p. 46).

This means that producers are using analytics to second guess the audience and the algorithms at the same time. This is not straight forward and causes headaches for journalists. For example, if a certain type of story told in a certain way provokes high numbers one week and much lower the next, is that because the work wasn't of the same quality this time, nothing more than bad luck or due to a change in an algorithm? All of which is to say that this is just one of the reasons why you must exercise caution when using analytics to inform your journalism. Speaking of which …

Critical issues

We'll finish this chapter by looking at how controversial analytics are in the field of journalism. Within the industry, knowing your audience through analytics, and creating content with them in mind, is largely seen as a virtue. The concern is what happens with the normative functions of journalism; holding power to account, informing the public about what's happening in the world and why it matters, the need to be factually accurate, exploring all perspectives of a story, etc. Does all that go in the bin in order to please the machine? Studies report that compromises are certainly being made: 'Data aggregation of audience clicks plays an intricate and dynamic role in influencing whether and how online newsroom editors decide to feature certain news stories' (Lee et al., 2012, p. 519).

This rings true with our industry specialists. Ryan from CBS Sports is frank about the level of influence, explaining, 'analytics is the be all and end all with social media. I work predominantly on YouTube, so that's where I'm getting all my analytics from and that is driving my decisions. My everyday decisions of what's gonna work and how I'm gonna work'. Laura, who's in social media marketing, is equally blunt, saying 'if you're not producing content that gets results then you're kind of failing at your job'. The influence of analytics on editorial decisions is also evident at the BBC, where Jacob notes that if 'content does well, we will just basically do more of that. It's only when you play the games of these third-party platforms that you've got the possibility of getting that high reach every time'.

In the race for more clicks, media producers are under pressure to tell stories that the public are interested in, at the risk of those in the public interest. To reach the audience, content is being made to fit the priorities of the platforms, putting a squeeze on news values.

But Jacob goes on to suggest that it is possible to harmonise between platform requirements and editorial goals, insisting 'you can't steer the editorial agenda based on how many engagements you got on Facebook'. He believes that studying audience consumption of your journalism can change the way you present a story, rather than necessarily changing the story you tell, and we've seen this in action in the section on social short videos. Jacob's not alone, as his thoughts are echoed by a digital editor at the BBC:

> We are no longer throwing our line in the river and seeing what hits it; we can to a certain extent, not predict, but we can see where we performed and didn't and improve that performance. That's not to say the story changes; it may be that our method of telling the story changes. (Bucknell, 2020, p. 12)

Digital producer Georgia from Channel 4 is also in agreement with the idea that you're still looking for the best story but might tell it in a different way depending on how the audience and algorithms react:

> The story that's the best one is probably not going to change. But it could be how you frame it changes depending on what the analytics are saying; whether they want shorts or long form content or keeping people around by sticking things right at the start.

You can use academic literature to further explore the benefits and drawbacks of using audience analytics to decide what stories are told and how we tell them. There is a lot to choose from, so these recommendations are just to get you started and you'll find the full references at the end of the chapter. Chris Anderson (2011) established how metrics influence news selection in online publishing and wrestled with what that means for the relationship between the reporter and the reader. Julie Firmstone (2024) offers a critical reflection on how analytics are changing journalism practices and the consequences for content. Kenza Lamot (2022) illustrates the impact of social media analytics on editorial decision-making by looking at how Belgian news organisations choose stories for Facebook, and the effect of audience metrics on practitioners is the focus of Caitlin Petre's (2018) ethnographic newsroom study.

References and recommended reading

Anderson, C. W. 2011. Between Creative and Quantified Audiences: Web Metrics and Changing Patterns of Newswork in Local US Newsrooms. *Journalism: Theory, Practice & Criticism*. **12**(5), pp. 550–566.

Barron, A. 2022. *Why metrics matter: Helpful analytics every journalist should know*. [Online]. [Accessed 15 July 2024]. Available from: <https://business-journalism.org/about-the-reynolds-center/>

Bean, A. 2021. *Crafting a video content strategy for social platforms*. [Online]. [Accessed 16 June 2022]. Available from: <https://www.poynter.org/reporting-editing/2021/video-content-strategy-social-platforms>

Bucknell I. 2020. Going digital, not dying out: how universities are uniquely placed to teach digital journalism survival skills. Journalism Education. 9 (1), pp. 6-17.

Christison, C. 2023. *How to Increase Social Media Engagement* [Free Calculator]. [Online]. [Accessed 15 July 2024]. Available from: <https://blog.hootsuite.com/social-media-engagement/>

Firmstone, J. 2024. *The shaping of news: a framework for analysis*. London: Palgrave Macmillan.

Harcup, T. 2015. *Journalism: principles and practice* (3rd edition.). London: SAGE.

Lamot, K. 2022. What the Metrics Say. The Softening of News on the Facebook Pages of Mainstream Media Outlets. *Digital Journalism*. **10**(4), pp. 517–536.

Lee, A. M., Lewis, S. C. and Powers, M. 2014. Audience Clicks and News Placement: A Study of Time-Lagged Influence in Online Journalism. *Communication Research.* **41**(4), pp. 505–530.

Newman, N. R, Fletcher, R., Robertson, C-T, Arguedas, A. R. and Nielsen, R. K. 2024. *Reuters Institute Digital News Report 2024.* Oxford: Reuters Institute for the Study of Journalism.

Petre, C. 2018. Engineering Consent: How the Design and Marketing of Newsroom Analytics Tools Rationalise Journalists' Labour. *Digital Journalism.* **6**(4), pp. 509–527.

Pew Research Center. 2024. *Social Media Fact Sheet.* [Online]. [Accessed 10 July 2024]. Available from: <https://www.pewresearch.org/internet/fact-sheet/social-media/#who-uses-each-social-media-platform>

Vogels, E. A., Gelles-Watnick, R. and Massarat, N. 2022. *Teens, Social Media and Technology.* [Online]. [Accessed 11 July 2024]. Available from: <https://www.pewresearch.org/internet/2022/08/10/teens-social-media-and-technology-2022/>

Journalism skills for the digital age

In the preceding chapters, you have learned how to produce journalism using digital media. We have studied the type of outputs that you'll be asked to create as a journalist, focusing on how stories are selected, told and disseminated in a digital environment. You have been guided by industry specialists and completed exercises that allow you to put your skills and understanding into practice. We have also talked about ways you can reflect on your experiences to get the most out of them. You're well on your way to becoming a digital journalist.

In this chapter, the focus is on the fundamental journalism skills and knowledge you need to be a digital reporter. We will cover how the core abilities that journalists have always needed apply when working with digital media, along with some new competencies that are particular to digital. They are skills that have been identified by digital editors who recruit and manage journalists for a range of news organisations. They include newspaper titles that are more than a hundred years old and digital startups that have been with us for just a few years. We also learn from graduates of our Journalism BA at the University of Leeds, who can bridge between studying the subject and their early careers as digital professionals. From their experience, they know what it takes to make it as a contemporary journalist and are well placed to guide you on what you need to learn before going into industry.

What is striking, and pleasing, from the point of view of people looking to break into digital journalism, is the amount of agreement that exists between this disparate group around the skills and understanding that matters. Although modern media production can feel turbulent and ill-defined, there are many common traits that can be deciphered from the chaos. In this chapter, you'll learn skills that you can be sure will help you through your journalism education, into industry and onwards through your career.

The section on *mobile filming* is necessarily long. Creating video journalism is a complex business, as you strive to capture and combine moving images, sounds, graphics and words to form a coherent communication. It takes a lot of practice, for instance, to get to grips with using a camera and essential accessories. Filming vertically for mobile consumption is also a relatively new discipline that we cover in depth with lots of illustrations. Plus, you'll learn how the stylings of digital video production differ from TV, in line with the expectations of social media audiences.

News sense

All the digital editors we spoke to identified the need for journalists to be curious about the world around them and to follow up on that curiosity by judging whether an event or circumstance should be reported on. No matter what the medium or what their role might be, journalists must know a story when they see one:

> Because no matter how digitally first and whizzbang-ly data driven and visually driven and mobile driven and socially driven our journalism is going to become, or indeed our audience will become, at the heart of a successful piece of journalism is always a great story. (Financial Times)

The digital-native vice.com agrees, saying that 'story is key still' and that it will always take precedence over 'whether it's written or a 360 live video'. It was noticeable throughout the interviews that the editors felt that a reporter's ability to capture the essence of a story could be demonstrated in their headline writing and the wording of social media updates.

Graduates of ours who now work as digital media professionals, gave us a valuable insight into the challenge of trying to identify stories when you're learning how to do journalism. As Laura Walcott explained: 'I struggled so much to find stories, like it literally was the most stressful part of the course'. Her classmate Jacob Tomlinson shared her feelings and said that this was largely due to a miscomprehension of expectations, as he thought at first that he needed to 'find like a Panorama or Newsnight exclusive'. He goes on to describe how he learned that it didn't need to be a Pulitzer Prize nomination, but a good story, of any scale on any topic, that told people something they didn't already know. Trusting your own thoughts, feelings and instincts is a good starting point. If you spot something that makes you think, feel or ask questions, there's a strong chance that your audience will feel the same way.

Exercise 6.1: Find a story and start developing it into a digital report

To help you develop confidence in your ability to spot a story, go for a walk with a course mate around the town or city you're based in and look for the best story you can find. Can you see anything that looks unusual or significant? It could be a row of shops where most have 'for sale' signs up, a group of campaigners asking people to sign a petition or someone creating a form of street art you haven't seen before. Talk to the people involved and ask them questions about what they're doing and why it matters.

Because our focus is on digital, think about how you could develop what you've found for one of the formats we've described in previous chapters. Complete the following three steps:

A one line description (the top line, as it's generally referred to in journalism) of what the story is and why people should be interested.

List three contributors you could interview in connection with the story, with one line each on why their perspective matters.

Say what digital form you would use to tell the story and on what platform, with a short justification of your choice and how it's a good fit for the available audience.

Interviewing

Sky News want journalists who can follow up on the curiosity that underpins their news sense by having 'an ability to ask questions based on that curiosity and have the confidence to do it in person or on the phone'. One way to approach the task is to get into character:

> If you're going to be a journalist in news, you're going to have to ask some awkward questions of people who don't want to answer your questions, and that might be people who are bereaved, or it might be slippery politicians, or it might be criminals. But your job is to ask those questions and so you need to develop a persona which will enable you to do that. (Sky News)

Having the self-belief to carry out interviews is particularly challenging for early career journalists. As BBC Yorkshire puts it, 'They need to be not scared to leave the office. I know this sounds stupid, but we see journalists increasingly who work in the office'. That's not surprising, given the amount of work that can be done remotely using digital tools for research, communication, production and distribution. A lot of your work as a digital journalist can be done from a desk. But to make the most of your contributors, get out of the office and talk to interviewees in-person, so that you can build trust and a rapport.

As with other topics covered in this book, we're not going to repeat learning that you've covered elsewhere on your Journalism programmes. You'll have been taught how to prepare for and carry out interviews, so what we can add is insight into how that plays out when you're producing stories for digital output. Jack Gilmartin is a Journalism graduate who now works in the digital department for Channel 4, mainly focused on sport and with responsibility for covering the England football team. His experience embodies the connection between the traditional skill and it's modern application:

> When you're making original content, you have to interview the players. I did an interview with Gareth Southgate (England manager) and it was good to have those journalism skills of asking open-ended questions. It's applicable to what we do, we're just doing it for a different purpose.

He explains that the digital team's approach to doing interviews is different to that taken by TV colleagues who work for the same organisation. Rather than going in with a preset list of questions that cover the main storylines before a game and asking them in an earnest way with a conventional interview setup, the digital creators work to 'get the best out of the players'. They aim to show the personality behind the performer:

> [We get them to] reveal something about themselves that they wouldn't have done otherwise, by getting away from the normal interview format. All the stuff we come up with, all the sort of mechanisms that we get within the content, is really just to get the players to relax and have a laugh and then we can clip that out for different platforms and it's really easy to use.

Jack gives the example of getting three England players to take each other on at a set of viral challenges, to bring out their competitive spirit and to show how they interact. His view is that this approach shows the audience what the players are like and how they feel in a way that interviews on their own cannot. The resulting video got half a million views on YouTube alone.

Being attuned to what the social audience is watching is key and Jack's team makes it their business to keep across trends, looking at audience analytics to see what's popular, and by following the accounts of influencers. Identifying potential interviewees, in this case footballers, who have a big social media following is another consideration and he cites winger Bukayo Saka as an example: 'We always push to get him at the forefront of our media just because he's big with the audience we want to target'. He's found that the England players are up for this social savvy approach because it's familiar to them and because 'they're good at marketing themselves, so they're prepared to throw themselves into stuff'.

The picture overall is that throughout the process, the long-established traits of good interviewing remain, but new techniques are being developed to bring them in line with the expectations and preferences of the digital audience and interviewees are being chosen because of their social media clout. You still want to get from your interviewees the vital and unique insights they bring to your story, but the lesson from our experts is that you must do it in a way that is relevant to the people and platforms you're targeting.

Summary

To make the most of your interviews for digital media, think about the following: talk to interviewees in person; ask open-ended questions; reveal the person behind the performer; look for contributors with big social media followings; build your confidence through practice.

Writing

The importance of this skill was also unanimously agreed upon and something that the digital editors were clearly passionate about. They stressed that no matter the medium, platform or format, being able to write is essential:

> Writing is number one for me. When we were recently recruiting, we had a lot of very talented people in video but struggling with the writing. We're after strong writers. If you can write well, you can write the article to accompany the video. You can sell it the best way you possibly can on social platforms. So, we definitely still need strong writers. (ITV News) (Bucknell, 2020, p. 14)

Without being prompted, our early career journalism alumni supported the views of their senior colleagues on written skills. We've already looked in some detail at writing online news but our specialists add that, within the sphere of digital, there are further approaches needed, depending on factors such as story, platform and audience. For example, in terms of platform-specific writing, with an Instagram Reel you effectively introduce the story at least twice, in the thumbnail and post. You may even do it a third time if the thumbnail is specially created, rather than defaulting to the first image in the video.

You want to avoid repeating yourself and we can see that in action in *figure 6.1*, with screen shots from an Instagram Reel by Channel 4 News. The first image shows the thumbnail. This element is also sometimes referred to as a 'preview' or 'cover', but whichever term you use, it's the image that users see when they scroll through your videos, deciding whether to watch them or not. In this example, large text declares the existential threat to a tribe. In the second, we see the opening caption of the video, that adds the important details that this is happening in India and relates to an indigenous group. The final screengrab is the most complete summary of the story, from the post that sits alongside the reel. It reads a lot like the opening line of an online news story and the structural relationship of these introductions is not unlike the inverted pyramid explained in the online news section, with the story, in essence, encapsulated three times with increasing amounts of detail.

Figure 6.1: Channel 4 News Instagram Reel with accompanying post on threat to a tribe in India.

Jack at Channel 4 was keen to underline the importance of the words used in thumbnails, as they're the audience's gateway to your content. He says they often use quotes in thumbnails and relates doing so to headline writing for online journalism, where a strong quote from a key contributor can catch the eye of readers.

Getting the right tone is also critical in writing for digital media. Georgia Carroll explains how the Channel 4 brand she works for in the UK encourages producers to write in a voice that suits their audience:

> E4 has been quite fun because they have quite a fun tone of voice in that way. So, they have this sort of character that they tell you to become, who's like this 25-year-old girl who uses social media and all this stuff. So, write in a personable way. Write in a way that you would normally speak rather than a company. That, I think, is more relatable for people watching it. (Georgia Carroll)

It's good to know that you, as young people starting out as journalists, have the advantage of already speaking the language of the audience you'll often be reporting for as a digital specialist.

Jacob Tomlinson, now at BBC Yorkshire, agrees on the need to match tone and audience, but says that even when writing for social media, there's a time for sticking to the more formal tones of traditional news writing. It depends on the story: 'We wouldn't write about the death of a much-loved person in our patch the same way we would write about the birth of baby Penguin in one of our zoos. You have to match the tone up with the tone of the story'.

Mobile filming

Another unanimously valued skill identified by the digital journalism professionals we spoke to is filming. This is hardly surprising coming from our broadcasters, but it was interesting to see that representatives from organisations with a publishing background agreed. A digital editor at JPI Media, a local and national newspaper group in the UK, explained how their reporters use mobile phones to gather video to either complement a written report or to produce a social short.

The use of mobile phones to film stories is nothing new; 'mojo', the popular shortening of the term 'mobile journalism', was first coined in the United States in 2005 by an executive at a news organisation in Florida that was pioneering the use of mobiles in reporting (Martyn, 2009, p. 209). The use of smartphones has gradually increased since then, boosted during the Coronavirus pandemic both in the industry and in universities, when sharing camera equipment became fraught because of the contact risks. Journalists became increasingly dependent on mobiles to conduct interviews and gather shots, even teaching contributors how to film images in their homes or workplaces to avoid breaking lockdown laws. Spanish television reporter Leonor Suárez explained to the Thomson Foundation that it was the key to staying on air (2021).

In the opinion of a JPI Media digital editor, whilst the recording device in use is much different to a professional camera, the video producer's role is not dissimilar: 'If you understand shooting, whether you're using a broadcast camera or a mobile phone, the principles remain the same'. As with TV news and any other form of factual television, you have to gather the images, sounds and interviews that tell the story. In this subchapter, we look in depth at how you should go about doing that whilst using your phone as a professional filming device. You'll also discover approaches to filming that are specific to digital journalism.

Setting up your smartphone

Let's start with how you should prepare your smartphone for use as a video recorder. You'll all have different mobiles with different cameras, microphones, apps and settings. However, certain controls, processes and preparations are universal.

- Clean your lens: It's not something smartphone users tend to do, but even a small speck of dust or grease on your lens is magnified when you display your work on a screen. So, get yourself a lens cloth. The kind you use to clean glasses are an ideal size.
- Do not disturb: The last thing you want when you're gathering shots or live streaming is your mum to call, asking when you're next coming home. So, put your smartphone into 'nighttime' or 'do not disturb' mode, as this stops calls and notifications coming through. Note that 'flight mode' would additionally prevent you from sending data and that makes live streaming impossible.
- Storage space: Video files are relatively big, and you don't want to run out of space to record, so now is a good time to have a clear out. Any media that you don't need to keep on the device itself can be safely copied to an alternative storage space, such as a cloud server or PC, before you delete it from your phone. Check if you also have any unused apps that can be deleted to make more room for your videos.

Choosing the correct type of video to record is important and the options shown in the settings for your phone's camera can be confusing. The choices you're offered will likely consist of the following three values:

- Format: In the settings for your camera app, you can choose the video format appropriate to your location. Broadly speaking, NTSC is used in North America, SECAM in Eastern Europe and China, with PAL covering much of the rest of the world, including the UK. Check online for the country that you're in.
- Framerate: Choose the framerate for your location. Again, North America is an outlier at 30fps (frames per second) with 25fps common throughout the rest of the world.
- Resolution: Higher definition video has greater resolution, meaning that the image is more detailed and sharper, particularly important when displaying on a large screen. If you're using your mobile to produce for TV therefore, 4K is a good choice. However, if you're making a social short, or any video likely to be viewed on a smartphone, 1080 (i.e. high definition) is fine and uses less than half the storage space of 4k, meaning you're less likely to run out of room.

All of this means that if you're filming in the UK, for example, you'd select 1080p at 25fps PAL for filming social videos. In the United States, you'd go for 4K at 30fps NTSC for filming for TV.

Now open the camera app on your phone and choose the video mode. There are lots of specialist camera apps available in the store for your phone, but the native app (the one that comes with your device) is fine for most mojo uses. Here are some further decisions to make with key functions that are common to many devices.

- Horizontal or vertical?: If you're certain that your work will only be used for social media and mobile distribution, consider filming the video with your phone camera in the vertical plane (i.e. portrait), as networks such as Instagram, TikTok and Snapchat only play videos in that orientation. If you're filming for TV as well, film horizontally and change the orientation in the edit.
- Focus: Your mobile camera automatically chooses what to focus on. This estimate is usually pretty good, but can let you down if, for example, there's an image in the foreground that you want to have in focus and it chooses the background instead (see *figure 6.2*). You can take manual control by tapping on the screen to select the part of the image that you want to have in focus. Lock in your selection by holding down your finger on the relevant portion. On iPhone, the message 'AE/AF lock' appears on the screen to confirm you've done this (see *figure 6.2*).

Figure 6.2: A camera automatically focused on a tree in the foreground and manual control is used to select the field as the focal point.

- Exposure: To manually adjust exposure (the brightness of the image), with your focus locked in, move your finger up to make the image brighter, down to make it darker. Tap outside the focus box to exit the lock and choose new settings.
- Zoom: You can also zoom in and out by spreading or pinching two fingers on the screen. Whatever device you're using, be aware of its optical zoom capabilities. For example, if you have a 2x optical zoom capable device, you can magnify the image to make it two times larger. If you go any further than this, the camera switches to digital zoom, which breaks down the pixels and reduces the quality of the image. It's best to stick within the range of your optical zoom and, if you really have to zoom in further, you can do this in the edit.
- Special features: If your app allows, you may want to use features such as slowmo (great for action shots in sport) or timelapse. Don't choose these options for the sake of it. Have a clear justification in mind, asking yourself what it adds to your storytelling.

Assembling your mojo kit

It's just about possible to shoot your news story using nothing more than your mobile phone, but from our experience and that of industry pros, you'll get much better results if you invest in at least a basic mojo kit. There's lots of guidance on what you might want to buy readily available from reliable sources on the internet, but this is our shortlist of the generic items that we see as essential:

- Hand grip: Holding your phone steady for long periods of time, without getting your fingers in the way, is near impossible. A range of hand grips are available that not only allow you to hold the lens more comfortably and with more control but also come with 'cold shoe' attachment points where you can add more kit in a secure fashion (see *figure 6.3*).
- Lapel mic: With phone cameras continually improving, often the biggest tell that something was shot on a mobile device is the sound quality, so it's worth investing in microphones. A lapel mic (posh name, 'lavalier mic') clips to you or your interviewee near the collar when talking on camera and ensures that the voice you want to hear is heard above anyone else or any background sound (*figure 6.3*).
- Onboard mic: Speaking of background sound, you also want to capture quality atmospheric audio when filming your shots and this can be lost if the wind's blowing against your phone's built-in mic. Windjammers that either slot over the end of your device or can be stuck over the mic specifically, are affordable and make a big difference. You could go a step further and buy a directional onboard microphone that can attach to your grip.
- Tripod: This would be next in line from our point of view (see *figure 6.3*). Lots of handshaking on the edge of the screen can detract from interviews and shots of static objects or scenes. You might get away with it for some stories, but there'll likely be times when it wouldn't suit the tone of the piece. Tripods for smartphones can be borrowed from your college or university if available and bought cheaply online if not.

Figures 6.3: Components of a mojo kit [Credit Samson Borthwick].

These few additions to your phone will make a big difference. There are many more gadgets you can invest in, depending on the type of work you're doing, but the four items listed above will get you off to a flying start. Note that you may have to buy adapters for wired connections to your mobile, most often a headphone jack to USB C adapter for both Android and iPhones.

Filming sequences

You have the mojo kit you need, your phone is set up to shoot and you're in manual control. We'll now look at the principles of filming journalistically that apply whether you're shooting anything from a short reel for social media to a documentary for TV. The fundamentals remain the same; you have to gather the images and interviews you need to show and tell people your story. That's pretty obvious, but knowing where to start, what to shoot and how many shots to gather can be daunting. This is particularly true if you're in a busy environment with lots going on or, at the other extreme, are filming a story that's picture challenged. But there are ways of going about the task that break it down into manageable chunks. We'll focus on your sequences of shots first and interviews second.

We advocate the *five-shots* approach to filming journalistic video. This involves identifying scenes within the environment you're in that relate to your story and for each relevant scene, shooting a range of five shots from different angles and of different sizes. These shots, or at least a selection of them, can be edited into sequences for your final piece.

So, for example, if you're filming a story about the hospitality trade recovering after economic problems, you might visit a restaurant that's seen a big increase in customers and is hiring more staff. You need to gather the pictures that tell that story. What scenes can you imagine being able to capture that reflect the restaurant's success? Well, a busy kitchen would be a good place to start and there's bound to be lots going on. You could get a sequence of five shots of somebody chopping vegetables in preparation for cooking, a chef working at a stove and the meals being brought to the pass for waiters to take to customers.

Once you've identified the scenes, it's a case of deciding which five shots you can record for each (see figure 6.4). For the preparation work, the main action is the chopping of the vegetables, so a close-up of a hand using a knife to cut into the food would be a good start. Next? We're human beings and we identify each other by our faces, so a close-up of the cook's face is important for the audience to see who's at work. You can now get some wider shots that put your close-ups into context. A medium shot from the side showing the person and what they're doing in one frame would be helpful. From there, an over-the-shoulder shot gives the cook's point of view looks good and has depth, with their head and shoulder in the foreground and the knife work in the background. Finally, a wide shot that reveals the prep work in the context of the whole kitchen, with other activities also in vision, would provide an establishing shot for your scene.

Figure 6.4: A sequence of shots of a sous chef chopping cucumber.

Now repeat. Many times. Five shots for your other kitchen scenes and the same for all the scenes at the front of house, such as a table of diners, the maitre d' welcoming guests and drinks being poured at the bar. Along with the activities you'd expect to see, keep on your toes and respond to any spontaneous actions, such as a birthday cake coming out to a table and people singing in celebration. You should also get a sequence of exterior shots, such as the restaurant amidst a row of businesses, the menu in the window and any outside tables. Look for images that reflect the story, like a sign saying, 'Bookings only, no walk-ins available', or 'Staff Wanted'. If you gather five shots of around eight scenes, you'll come away with forty images and, along with your interviews, that's ample for a digital video.

Be aware that you won't use all your pictures as you won't have space. You might in the final edit, for example, only use two shots of the veg preparation and that's fine; it's always better to have too many than too few. Also note that you don't have to use sequence shots in the order you filmed them. The over-the-shoulder of the chopping might look great, showing the audience what's happening in one image, so start with that, then edit into the close-up of the face. The pictures of diners eating might be the last you gather but work well at the beginning of the video to show how busy the restaurant is.

The principles you apply to filming people at work extend to all other actions and scenes. For example, if you want to capture someone walking:

- get a close-up tracking shot of their feet (i.e. stoop behind them with the camera near ground level for a few strides);

- a tracking close-up of their face;
- a tracking over-the-shoulder seeing where they're going from their point of view;
- a medium shot with the phone held still as they walk through the frame;
- a wide silhouette image from the side with the sun in the background.

Shooting 6–8 scenes for 30–40 sequence shots is a good target to aim for; more than enough for your video and not so many that you spend too much time filming and create the headache of having an overabundance of pictures to edit. Use the *five-shots* approach to create a shopping list of images in advance that you need to gather on location and tick them off as you go. Remember to be flexible and gather additional scenes that you didn't anticipate but connect to your story. They'll likely replace things you thought you'd see that don't happen. But going in with the list to start with gives you confidence that you'll get enough shots. If you struggle to imagine what you'll be able to film, it's a warning that your story might not work well for video.

Exercise 6.2: Practice shooting sequences and make a shot list for a news video location

There's a couple of things we'd like you to do to practise your sequence filming. The first is to bother your housemates, classmates or family and shoot sequences of them doing something. It could be anything and you can start modest. Capture them preparing a meal, going out to the shops or sat around talking. The more you film the more comfortable you'll become with putting the *five shot* theory into practice in a range of scenarios.

The second is a written exercise. Work out how you might film a news-related event, activity or location. It could be a protest, performance, sports training, transport hub, street scene or one that you come up with yourselves from your experience of making and watching news. List six scenes that you can anticipate finding and for each describe five shots you could take. Apply the terminology we have used to describe the size of shot, the subject and, potentially, the angle. A shot list for protestors gathered in a city centre might look like this:

- High angle wide shot of crowd.
- Medium shot (i.e. whole subject fills frame) group of protestors at ground level.
- Low angle medium shot of a placard.
- Close up protestor chanting.
- Over-the-shoulder of same protestor for their point of view.

Further scenes could be a speaker addressing the crowd, by passers watching on, the group starting to march through the city, police/stewards walking with protestors and the break-up of the demonstration at the end.

Filming interviews

As a mobile journalist, filming with a compact kit and likely by yourself, you need a quick checklist for ensuring that your interview shots are set up correctly. So here it is!

- Audio: This is where the lapel mic we suggest you invest in comes into its own. With a mic clipped to their collar, you'll cleanly pick up your interviewee's words over background noise. Be sure to test the mic after you've plugged it into your device, to make sure the camera app is picking it up. Do so by recording a test clip and scratching the tip of the mic and/or blowing on it. When you play it back you should hear the distortions and you know you're in business. If the sound is clear, it suggests the app is using the smartphone's internal mic. Check the connections between your wires and the camera's microphone settings if you have issues.
- Composition: The convention in TV journalism is for the interviewee to be framed according to the rule of thirds style of composition. This is where the subject of an image is off-centre, a third of the way in from the top or bottom and one of the sides of the frame. When you're filming vertically for digital platforms, the side element is taken out and the aim is therefore to position the shot so that the subject is a third of the way from the top or bottom of the screen. In the case of interviews, you should use the eyes of the interviewee as a guide and move the lens so that they're a third away from the top. If you have the contributor sat lower in the frame, you get a lot of empty head space and the image looks imbalanced.
- Eyeline: The next consideration is where the interviewee will face. In TV, the standard is that they look off-camera towards the interviewer (*Figure 6.5*) who's sat to the side of the lens. This works for digital video too but is less strictly adhered to with interviewees often looking straight to camera (*Figure 6.5*). This style works well because viewers are used to content creators looking directly at them on social media and suits the intimacy and informality of many platforms.
- Eye level: It's deeply unflattering to point the lens from a low angle up someone's nose and equally so with a high shot that squashes their facial features and exaggerates the size of their foreheads. Do everyone a favour and adjust the height of your camera on the tripod so that the lens is approximately level with their eyes.
- Background: What's behind the interviewee matters too. There's no single correct approach as it depends on the style you want for your video. A general rule is not to sit them with their back to a blank wall because this creates a flat background with no visual interest. At the opposite end of the scale, you don't

want too much action going on behind the contributor as it could distract viewers. So, generally, look for a background that is a distance behind the interviewee (for depth), is relevant to the story, but not too busy. For example, if you're interviewing a band member, position them on the side of an empty stage with instruments behind them.

- Light: The term used in videography for the brightness of the image is exposure. To expose the interviewee correctly, position them so that the dominant light in your environment shines on their face (*figure 6.5*). When the main light source is behind your subject (i.e. 'backlit') their face is under-exposed (see *figure 6.5*). If you're outside, have the sun at your back and, if it's cloudy, check where the sun is by holding up your hand and seeing which side is in shade and which side has some light on it. Indoors, it depends on the available artificial lights but so long as one of them illuminates your interviewee's face, you're in business.

Figure 6.5: Setting up interview shots correctly.

All of this sounds like a lot of work now we've written it down and explained it but many of these attributes can be achieved in how you set up the interview before calling your guest into position. If possible, conduct the interview sat down with two chairs directly facing each other (see *figure 6.6*). If you're outside, a bench or low wall provides a similar scenario. Place the camera next to your seat, so that you can readily check the image, and get it to eye height by adjusting the tripod. Make sure the interviewee's chair is facing the dominant light source and isn't up against a wall. This 'set' ensures that all the visual components of the interview picture can be achieved with a few small changes once the interviewee is sat down. Help your contributor to clip on the lapel mic that you've already tested and away you go!

Figure 6.6: The two-chair set up for an interview shot vertically for digital video.

Shooting for social

There are additional traits of mobile filming that are particular to producing the kind of digital videos we've focused on in this book. BBC social media reporter Jacob Tomlinson explains that key to your approach to filming for digital media is audience expectation. Videos that users are used to seeing in their social media feeds are made by members of the public, who are not aspiring to the production standards of TV professionals. Jack Gilmartin, a digital content creator for Channel 4, explains, 'it's OK to get stuff out that's not absolutely polished to perfection'. As we've touched upon elsewhere, journalists look out of place if they turn up to informal social spaces, real or virtual, wearing a buttoned-up shirt, tie, suit and shiny shoes. You can drop the formalities and allow your shooting to be a little rough around the edges.

Jacob keeps recording from the point he starts working with a contributor because he knows that insightful, humorous or visually appealing material can be captured in the moments between shots or questions. He says BBC News has found that 'some of the best performing videos on TikTok, Instagram and Facebook, have been shot with a mobile phone', using the more rugged approach that Jacob encourages. For him, people on social media do not care about filming conventions, asserting that the audience isn't going to switch off if, for example, you do a zoom part way through an interview. He acknowledges that it's hard for camera crews who've been taught the conventions of TV news filming to adapt, but as upcoming digital journalists, you don't carry this baggage.

CAUTION! This is not to say that mobile filming for social should be done without thought or preparation. You can approach your work in a chilled-out way, but you still need to gather the shots and interview clips required to tell the story and that takes planning. Have a vision in mind for your social video before you set off to film, storyboard your ideas, think about the questions you'll ask and the images you need. Writing down all those elements is the best way to go until you're confident that you can hold a complete plan in your head and it's not unusual for video reporters to carry on with this approach, in some form, throughout their careers. There's a lot to think about when you're telling a story using pictures, sounds, words, graphics, music, etc. So, we're advocating disciplined effortlessness!

Video editing

Once you've gathered your pictures and interviews, the next step is to identify the best bits and craft them into a story suitable for digital distribution and for this you need to be able to do some video editing. This is a skill that pre-digital was carried out by craft picture editors who specialised in the role. Such people still exist and work on higher budget or longer-term projects, but in digital news production, it's more common for journalists to do their own editing.

That's been Jack's experience at Channel 4:

> Most producers now edit their own stuff because you're not having to go to that really high level that a professional editor goes to. If you can just knock something together and make it watchable, do a bit of a sound mix, you become way more valuable, especially if you're working in a fast turnaround environment.

Traditional TV editors will be spinning in their edit suites at this idea, but it's the reality of current and future video journalism and in keeping with what our specialists have said about a more informal approach to filming; keep it casual to suit the platforms and their audiences.

We've already guided you through the process of editing social short videos in Chapter 1 on digital video creation, so we won't repeat those lessons here. What we can add at this point are some editing techniques that our experts have identified as being important to all forms of digital video production, whether that's a social short, TikTok, explainer or video embedded into an online news story.

Ryan Daniels, social media producer at CBS Sports, says that the captions you add during editing not only perform the role of telling the story for the many users who don't listen to the audio that comes with video, but are key to retaining attention because of their visual impact. He advocates dropping formal, news-style captions with Arial font and being more creative with your text, having fun with it. He likes 'big pop art, colourful and engaging subtitles that are just popping onto the screen. You're immediately like 'ooh, these are cool. I'm gonna keep watching'. The left image in *figure 6.7* is a frame from an ITV News TikTok that flashes words onto the screen one at a time. The text in the right image is more conventional in length, but the font is an informal, handwritten style. Both suit the levity with which the story is told by the presenter.

Figure 6.7: Examples of single-word captions and stickers An ITV News report and an LA Times story with single-word captions published on TikTok.

Ryan acknowledges that an entertainment-style approach won't suit every story, but we've discussed elsewhere how younger audiences want even serious news to be treated with levity where possible. That's the case with the LA Times report above where the editor has included playful stickers and freehand-style text that matches the dry humour of the presenter.

Jacob at the BBC says that with interviews it's not just the filming that can differ from TV conventions when you're making digital content. He is more experimental and less conventional in the way that he edits interviews, again to better fit with social media stylings:

> You can do simple things like jump cuts. If you've got an interviewee who is saying something really emotive, I might, for example, on the emotive bit, just blow the frame in (i.e. a fast zoom). If your eyes see something happen in frame, you engage for a bit longer and so just that simple jump helps. (Jacob Tomlinson)

Laura Walcott is a journalism graduate who now deploys the skills she learned on her programme working in social media marketing. She extols the virtues of using music in social videos and making sure that you choose tracks that suit the platform and audience: 'I'm not gonna put any classical on there'. She devotes time to editing cuts and movements within the frame to the beat of 'music that is trending or is going to catch people's attention'. Laura says that her analytics show that videos made with music generate more engagement than those where the audio is limited to voice.

Social listening

A new journalism skill that feeds into areas of production that we've looked at is social listening. In varying ways, the Financial Times (FT), BBC Yorkshire, ITV News, vice.com and JPI Media, described the importance of this activity. Through different approaches, they tap into comments and social media communications to develop story ideas and editorial strategy and form a closer relationship with their audience. The FT employs what it calls a 'community manager' whose specific role is to build engagement through comments:

> We consider commenting, and the conversations that take place in comments, as being part of our journalism and increasingly most news publishers are. And there are ways in which we can even use those communities to feed our journalism. This is a whole subgenre of journalism which didn't exist before comments. (Bucknell, 2020, p. 11)

This echoes a study by Chen and Pain that describes how journalists have warmed to the idea of interacting with online comments as a way of 'fostering mutually beneficial connections with the audience' (2017, p. 876).

BBC Yorkshire describes a similar practice for harnessing views and experiences shared on social networks:

> We do lots about social listening, which you could say is harvesting content from social media so we don't have to get off our arses. I'd say it's different; I'd say it's harder to get to the pub these days so those conversations don't happen there anymore. In local news groups, people do speak freely and tell amazing stories without even knowing it. (Bucknell, 2020, p. 11)

This digital editor goes on to give an account of a story developed from a Facebook group called Leeds Face, where someone posted about their dad passing his driving test for the first time at the age of 79, so he could drive his wife, who had cancer, to hospital appointments. As ITV News put it: 'we often do pick up stories that are making waves on social media but haven't yet crossed over into news'. Embracing digital audience interaction and then responding to it in a way that informs editorial output, is one clear addition to your reporting tool bag.

Exercise 6.3: Practise your social listening skills

Time for you to put your social listening skills to the test. There are lots of different ways to scan what people are talking about on social media. Here are just three ways of going about the task of finding stories and ideas by listening in.

Search out a local Facebook group for your area and comb the posts for something that could potentially become a news story. You'll have to sift through loads of people complaining about bin collections, dog poo and bus times – but there might be something there that genuinely stands out.

Look at the comments that people leave on interesting posts on social media or on an online news story. See if there are messages that offer further insight or contradict the published narrative in a way that could lead you to an original story of your own. Are there gems of real-life experience you could follow up?

You can see what people are interested in by exploring trending topics on social media. Take a look at individual platforms or at Google Trends to find the search terms, hashtags and key words that are hot at the moment. Use these as a starting point for your investigation to find a fresh story.

Data visualisation

In social storytelling, Chapter 2, we highlighted 'social cards' as a way of reporting in a compact form that suits many social channels. This practice of telling stories through graphics that clearly communicate facts and figures is called 'data visualisation'. It's not particular to journalism but, as we see with social cards, it can be effective at conveying stories where there is strong statistical evidence or to complement other storytelling forms, such as a video package or a written article.

Creating graphical stories is something that people in the expert role of 'data journalist' specialise in; analysing data sets to find patterns, anomalies and any other features that indicate a phenomenon worthy of investigation. Data journalists often work with the assistance of data analysts and graphic designers but, depending on the scale of the challenge and the budget of their employer, sometimes independently. Visualisations can range in scale from simple graphs to interactive projects where the audience can perform searches or adjust parameters to see changes.

The advent of the internet has led to an age where more and more data is available to the public, and more easily. However, very few people have the skills or time to be able to check through it. The job of a data journalist is to spot significant trends or anomalies in the figures and let people know what is happening and why it matters.

Production

Although data journalism is a specialist role that is not core to digital journalism, with a bit of guidance and some recommendations you can soon get to the point where you're able to create simple and effective graphics that illustrate data to your audience in an engaging way.

Understanding data and interpreting figures may not be something that you consider your strength. However, many news stories rely on data to demonstrate or prove a point so being able to show this in a simple way is a key skill. It's important to recognise that there can be issues in the way that data is used and presented. Asking some simple questions will help you to avoid obvious pitfalls.

Where does your data come from? Firstly, is it public data from a trustworthy and reliable source? If it's a survey how big is the sample size? Who have they spoken to? A survey asking twenty teenage boys what they think of an issue could not be said to truly represent public opinion.

Are you getting the whole picture? You may have heard the phrase 'lies, damn lies and statistics' suggesting that data can be interpreted in many different ways and it's often possible to find numbers that can be used to back up an argument on any side. It's important to check that the figures you are using are not produced by someone with an agenda, always go back to the original data source to check you have the whole picture.

Are you comparing like for like? A big change in numbers could indicate a potential news story. However, the difference could come from a change in the way the numbers are collected, rather than a change in the issue you are measuring. For example, a reclassification of the term 'domestic violence' or a campaign by police to encourage more people to come forward if affected by domestic violence may lead to a rise in the number of reported cases, but not necessarily the number of incidents. Comparing cinema tickets sales in 2019 and 2020 will show a drop, but it is likely to be because of pandemic restrictions rather than a lack of interest in movies. Check any trends seen in data with the original source provider or an expert in the field.

Am I trying to find patterns that aren't there? Remember that correlation does not equal causation. Just because two sets of data show a similar rise does not mean that one has caused the other. For example, in summer months the sales of ice cream rise. In summer the number of shark attacks also rises. This is because the hot weather means more people are eating ice cream and more people are swimming in the sea. It does not indicate that eating ice cream causes shark attacks.

Data visualisation apps

Once you have found and checked your data, there are many free applications available online that will allow you to make a graphic or animation with relatively little hassle.

At the simplest level, both *Excel and Google Sheets* will allow you to create a chart within a spreadsheet. This will lead to a very basic graph to help the audience understand your data. In Microsoft Excel, you simply highlight the data you want to use in the spreadsheet (including any labels) and then click 'Insert' and 'Recommended charts' to choose from a variety of graph styles.

Data GIF Maker, by Google News Lab, has been developed with journalists in mind and intuitively leads you through a short production process in the creation of an animated data graphic suitable for sharing on social media. You can save your work as a jpeg file or copy an embedded link. Its beauty is its simplicity; even the most technophobic will quickly become comfortable with how to use the tool. But you might quickly get frustrated with the limited options available. For instance, there are only three themes to choose from to present your data and few alternative colours. To keep it suitable for a GIF, you can only represent four values and the title can have no more than twenty-eight characters, which makes life tricky because there's no other textual option available to explain the data you're presenting. However, as *figure 6.8* illustrates, it's effective at creating a quick and clear visualisation of small data sets. The Office for National Statistics data on the hit to business and tourism visits to the UK, caused by the coronavirus pandemic, is clear to see.

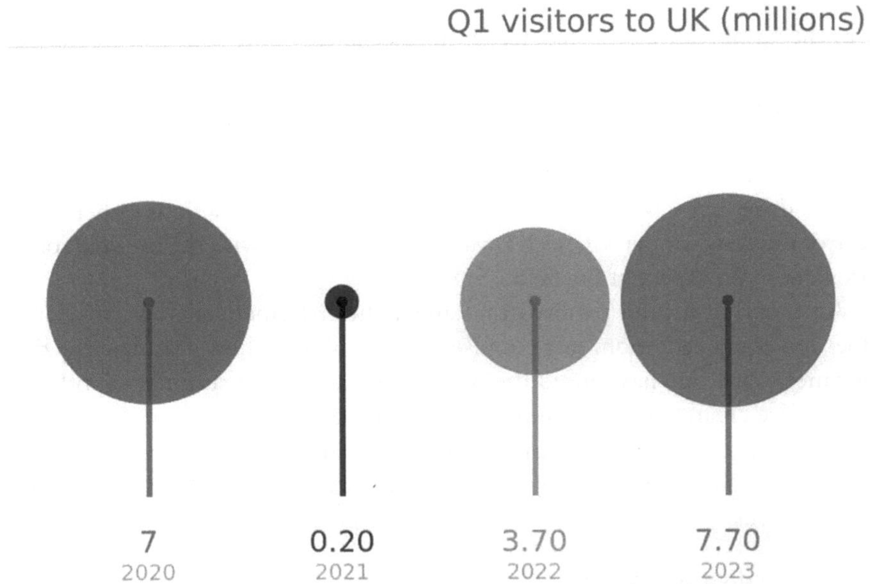

Figure 6.8: A basic data visualisation using GIF Maker showing tourism to the UK recovering after the coronavirus pandemic.

You can sign up to *Canva's* graphic design software for free. It's a popular system used by leading news producers, such as the New York Times, and is a big step on from GIF Maker in terms of the options available to you. Therefore, it takes more time to become familiar with the programme. But the support tools are sympathetic and include a tour of the application's functions, starting with the impressive range of templates that are organised by categories such as 'Infographic' and 'Data'. Click on the latter option and subcategories appear that further refine your search for the right template and include 'Statistics', 'Data analytics' and 'Charts', all useful for our purpose of data visualisation. You can customise the templates by removing elements you do not want or adding your own images, colour schemes, text and drawings.

Within the application, you can save your work as you go so that you can finish a piece off later or use it as a template in the future. Once complete, you can download your data visualisation or instantly share it on social media. Usefully, the default dimensions of your Canva graphics suit the shape of mobile phones, now the most common device for accessing news. You can download the Canva app to your mobile and work on the go or use it on your desktop with the benefit of having more room to manoeuvre.

Figure 6.9: Same data as previous illustration but with extra information, produced using Canva's infographic maker.

The templates are quite sophisticated with lots of layers of graphical and textual detail. Unless you happen to find one that exactly matches your needs, you'll have to modify what you have chosen to make sure it works for you. Happily, the menu items and options are intuitive to navigate and with some trial-and-error experimentation, even the mature authors of this book could quickly get to grips with making an infographic that included data visualisation. If you get stuck, alongside the in-app support, there are plentiful guides to using Canva's infographic maker available on YouTube.

Canva also has tools that enable you to automatically transform your data into a graphical design. Just head for the 'Elements' menu on the left of the screen and choose 'Charts' to explore the various line, pie and bar diagrams that are readily available. You'll be prompted to add your labels and input your data. You can even import data directly from a spreadsheet or CSV file, though you may need to refine the results, depending on how the system interprets the sheets. Alternatively, you could create your own image using the available shape and image tools, though you may have to get your calculator out to figure how big the shapes should be to accurately reflect your data.

Adobe is getting in on the act with an all-in-one content creation app that, amongst other things, enables you to produce graphics such as data visualisations that are suitable for social media apps and smartphone viewing. The package, called Express, launched in 2023 and was initially free but currently requires a subscription, either to the individual app or as part of Adobe's Creative Cloud.

Other free and easy online tools are available, such as *Datawrapper* or *Flourish*. These allow you to choose from a range of different charts and maps and produce effective and simple data visualisations with just a few clicks.

For a more advanced option, Tableau is a highly regarded and long-established player in the data software market, founded in 2003. They sell sophisticated data analysis and visualisation tools capable of powering large multinational corporations. As a student of journalism, that is a lot more punch than you need or would want to pay for. Thankfully, they also offer a free version to download called *Tableau Public*. They supply lots of 'how-to' videos on their website and helpful YouTubers have also been busy creating beginner's guides to this version. Even this slimmed down edition of Tableau is primarily designed to enable businesses to analyse big datasets, so it can take a while to pick through the help videos to find what you might need as a journalist.

Exercise 6.4: Create a data visualisation that functions as a news report

Using one of the free graphic design applications we've suggested, or one of your own, have a go at telling a story by presenting data in a graphic. You can find tons of data sets on the USA-based Kaggle website, designed to help people experiment with data analysis and processing. In the UK, the Office for National Statistics is a comprehensive and wide-ranging resource, providing data on subjects including crime, elections, leisure activities and well-being. From their website you can browse within such categories to find a data set that interests you.

The ONS publishes reports, or 'releases' as they call them, that highlight what they believe are key trends. You should take note of them but also dig deeper by opening the links they provide to spreadsheets that share the raw data, allowing you to investigate for yourself. Try to find a trend or phenomenon that says something the public would be interested to know. Look for changes over time, regional variations and demographic differences, trusting that if you spot something that stops you in your tracks, it's likely your audience will feel the same way. Once you have the data, edit it for the purposes of your story and begin the import->create->share process.

To sum up, we have explored the market for you and identified programmes with varying levels of accessibility and complexity. Depending on your skill level and your interest you may want to stick to the simplest or you may want to explore further.

We recommend that, if you develop a passion for data journalism and visualisation, you further explore the subject outside of our text using the wealth of available resources. Paul Bradshaw has researched, written and published extensively on data journalism, leading a masters programme solely focused on the activity at Birmingham University in the UK and acting as a consultant and trainer to news organisations such as the BBC, Sky and Trinity Mirror. For a critical take on the use of data by journalists, Professor Chris Anderson's *Apostles of certainty: data journalism and the politics of doubt* is a seminal text, exploring the history of how journalists use quantitative data in their reporting and how it might contribute to countering the lack of trust in professional journalism.

Graphic design

Outside of data visualisation, the journalism alumni who we spoke to for this book downplay the need to learn broader graphic design skills at university. They told us that although they frequently use graphics in their productions, they rely on professional designers for more complex work and have taught themselves how to use the tools to create basic images. It's a reminder not to sweat the small stuff and it seems in digital journalism, that stuff includes specific software and even certain skill sets.

Ryan from CBS Sports describes *Photoshop*, for example, as being user-friendly and explains that journalism graduates have a head start familiarising themselves with such an application because, 'we're all in the creative industries, we all have an eye for what looks good'. His former classmate Georgia, now a digital producer for E4, explains how she built up her graphic design skills and confidence by using the tools within a familiar social media platform:

> I used to use Snapchat because it actually has quite good editing tools within the app, particularly the sticker function. I just used to make really bad memes, like poorly handmade memes, and you can use scissors and cut things out on Snapchat and just work around that and be creative. Then I started doing Photoshop instead and kind of teaching it to myself. (Georgia Carroll)

Laura now works in social media marketing and advises that at the outset you should brace yourself for making mistakes and not to be discouraged. She says she, 'learned on the job, all self-taught. And yeah, you do learn as you go along. It can be pretty shocking at first'. Ryan sums up where he believes these particular skills lie in the pecking order: 'If I was hiring a graduate straight out of uni or anywhere, I would be happy if they didn't have any graphics design skills as we've all learned on the job'. For him, it's more about character than technical credentials, with a passion for social media being the number one requirement if you want to get into social journalism.

Along with the tools available within social media, the *Canva* infographic software we highlighted in the previous section is also user-friendly for beginner graphic designers. You'll quickly learn the options available to you for creating backgrounds, shapes, textual elements, charts and logos, with the option to import images of your own, such as photographs. The Canva desktop provides you with generous workspace when you use it on a computer and that helps with being able to find the key options.

Frost et al make further helpful software recommendations for getting started with graphic design. *Inkscape* is open source and good for all sorts of technical illustrations. *Illustrator* and *InDesign* are part of the Adobe Creative Cloud bundle and *PowerPoint*, part of the Microsoft Office suite, is also recommended as a way of making basic data graphics with an application that many of you'll have access to and be familiar with. Official video tutorials for these applications are provided by the publishers and unofficial guides and tips are available in abundance on YouTube.

Fact checking

Verification and fact checking are essential journalism skills that need to be applied rigorously in an online environment. In the immediate wake of the Manchester Arena bombing (2017), several social media users posted pictures they claimed were of relatives attending the concert where the bomb went off, who were now missing and potential victims of the attack. Many of these were shared by well-meaning members of the public and even reported by professional media. But the claims and pictures were false.

Spreading lies and false information is by no means a new thing. However, the twenty-first century has seen the weaponisation of information on an unprecedented scale (Ireton and Posseti, 2018). Digital and social media make it easier and quicker to publish falsehoods and disseminate to a potentially huge audience. The way social media is designed to allow people to consume bite-sized pieces of content and share at speed does not promote the provision of full context or encourage further research or detailed checking.

The concept of 'fake news' is one that audiences are well aware of, although it's a phrase that's come to have a complicated meaning. In an online context, it was used in 2016 to describe a cluster of websites set up in the Balkans. The pro-Trump sites were not created for editorial or political reasons, but to monetise Facebook. Writing sensationalised or false news stories with headlines such as 'Pope Francis Forbids Catholics from Voting for Hilary' and then sharing these stories on Facebook encouraged users to click (see clickbait) and so generated advertising revenue. But over time the phrase was adopted – perhaps most memorably by Donald Trump himself – not to refer to this fictionalised news, but rather to cast doubt on genuine media and reporting. It has since become such an over-used phrase, used to describe everything from misinformation to political spin, to mistakes or just stories that people don't agree with.

The development of technologies such as image manipulation and AI make it easier to fake convincingly. Generative AI is one recent development that has aided the spread of disinformation and misinformation. It can be used to manipulate or create realistic images, videos and audio. These 'deepfakes' can make it look like a person is saying or doing something that they have not.

As a journalist, you'll always want to check your facts correctly and tell the truth. Getting something wrong will impact negatively not only on the audience's view of the story you're telling, but also your own reputation, that of the organisation you work for – and potentially contribute towards lack of trust in the media as a whole.

All this leads to a situation where the public does not know what to believe. An annual international survey of digital news users found that in 2024 around six in ten (59 %) say they are worried about what is real and what is not online. The figure is considerably higher in South Africa (81 %) and the United States (72 %), both countries holding elections that year (Newman et al., 2024).

The same survey found the concerns about trustworthy and untrustworthy content is highest for TikTok and X when compared with other online networks. Both platforms have hosted misinformation or conspiracies around stories such as the Princess of Wales' health, as well as AI-generated pictures and videos.

You need to be aware of the ways that information can be used to manipulate or mislead. Let's start by looking at some basic definition of the kinds of false information that are out there. These will help you think about how and why information could be used.

- *Misinformation* is information that is false, but the person who is sharing it believes that it's true.
- *Disinformation* is information that is false, and the person who is disseminating it knows that it's false. It's a deliberate and intentional lie told to cause trouble or influence people.
- *Malinformation* is information that is based on reality, but used maliciously to inflict harm on a person, organisation or country. This could include taking facts out of context or selecting only specific information from a wider dataset.

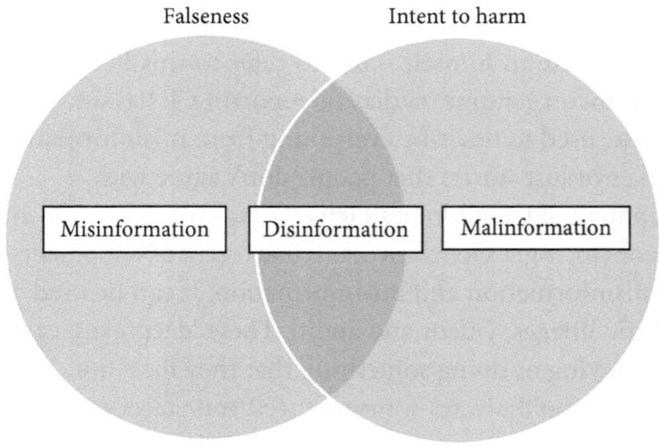

Figure 6.10: Information disorder (Wardle and Derakshan, 2017).

Within these types of information disorder, it's possible to break down seven different types of false information (Wardle and Derakshan, 2017). These help us to understand where information can come from and its original purpose.

- Satire or parody: Content that has no intention to harm, but potential to fool people into think it's real (for example something said by an impressionist in a comedy programme later being attributed to the actual politician they were impersonating).
- False connect: Where a headline, image or caption does not support the content (an example would be a picture of a celebrity used to attract attention to an article, where they actually have little or no connection to the story).
- Misleading content: Using information in a misleading way (an example may be only including part of a quote; or using one statistic from a whole report that supports your story, but not the ones that don't).
- False context: Genuine content shared in a false way (an example would be using a picture taken many years ago and claiming it was taken recently).
- Imposter content: Where genuine sources are impersonated (for example a false news story shared using a respected broadcaster or organisation's logo/design so it looks as though it's come from them).
- Manipulated content: Genuine content is altered in some way in order to deceive. This is most often video or images, for example altering an image to make it look like it was taken elsewhere.
- Fabricated content: Completely fake information (for example, a story that is entirely made up, or a fully computer-generated image).

False information can originate from anywhere – government sources, campaigners, celebrities or members of the public. It could be originated by a genuine mistake or created on purpose for financial gain or to cause harm.

If you do not check your sources and facts carefully before publication you run the risk of falling victim to a hoax, further spreading false and potentially damaging information or adding legitimacy or a wider audience to someone else's agenda.

What is being done?

Concerns over misinformation and disinformation became particularly apparent during the Covid-19 pandemic where people were living in isolation and seclusion and relying on social media for connection and information. This ranged from false claims blaming certain communities for spreading the virus to the reposting of incorrect medical advice, e.g. that the heat in saunas can prevent Covid-19 (Brennen et al., 2020). A number of measures were put into place by social media platforms to flag up concerns to readers or remove posts that break platform rules (e.g. Meta's Third-Party Fact-Checking Program). On most social media platforms, it's now easy to report posts for false information or misinformation.

We've also seen a rise in 'fact checking' and 'myth busting' as a journalistic discipline. Many media companies have introduced specialist units (for example the BBC Verify) or journalists whose role is to find and investigate misinformation and disinformation. Reputable news organisations will produce content such as 'explainers' (see Chapter 1) to present facts to audiences in a way that allows them to easily understand the truth behind a story.

There are an increasing number of websites dedicated to fact checking and pointing readers to legitimate research (such as FactCheck.org, snopes.com). There are also websites dedicated to media basis so audiences can better understand any motivation behind the way information is presented.

Exercise 6.5: Analyse a fact checking process

Take a look at some of the media fact checking services. Here are some examples you could look at:

Reuters Fact Check
BBC Verify
Channel 4 news FactCheck
Snopes
FactCheck.org
AFP Fact Check

These services have many examples investigating stories or claims that are being circulated online. There may be some that you have seen yourself while scrolling through social media.

Take some time to see how the fact checking is carried out for any story that catches your interest. Would you have known at first glance whether the story was true or not? What research did they do? How did they confirm or deny the claim?

What can you do?

As a journalist you can take steps to protect against information disorder.

First of all – check your own work thoroughly. Be sure to check any facts, figures or quotes with primary sources rather than relying on second or third (or more) hand information. If you need to, take the old-fashioned route and call someone up to ask them to verify what you're saying.

Don't make assumptions, treat everything with some level of scepticism. Before you repeat a fact or repost a story, think about who stands to gain from a particular piece of information and why it might be being shared. This is a list of questions you can ask yourself, drawn up by First Draft news (Wardle, 2019).

- Provenance. Are you looking at the original account, article or piece of content?
- Source. Who created the account or article, or captured the original piece of content?
- Date. When was it created?
- Location. Where was the account established, website created or piece of content captured
- Motivation. Why was the account established, website created or the piece of content captured?

There are a number of verification tools available online that can help you to check content, many of which are free to use. In Google you can carry out a reverse image search – using a picture as the start point for a simple search you'll be able to see if it has been used in any other context before. TinEye is another reverse image search tool that will find where an image has been used before. Image forensics tools will enable you to check an images metadata and see when it was created and whether it has been manipulated in any way.

There are several AI tools that are able to help detect AI-generated content. Different social media platforms have ways of verifying accounts and content, for example YouTube DataViewer enables you to see more information about a YouTube video and the account that created it, including time, date and modifications. These are being developed and changing all the time, so it's worth looking to see what is out there.

You should report anything that you see online that is not true. Most social media platforms allow you to report individual posts. If you see content that is harmful there will be official channels to report that, or, if necessary, report it to the police.

If you do use AI or generative technologies, make sure you label that content clearly so people know how it has been produced.

As a journalist you can also pro-actively work against misinformation by investigating any new cases of misinformation or malinformation you find. You can promote fact checking and myth-busting services that will help audiences reach the truth.

References and recommended reading

Anderson, C. W. 2018. *Apostles of certainty: data journalism and the politics of doubt.* Oxford University Press.

Anderson, C. W. 2011. Between Creative and Quantified Audiences: Web Metrics and Changing Patterns of Newswork in Local US Newsrooms. *Journalism: Theory, Practice & Criticism.* **12**(5), pp. 550–566.

Brennen, J. S, Simon, F., Howard, P. N. and Nielsen, R. K. et al. 2020. *Types, sources, and claims of COVID-19 misinformation.* Oxford: Reuters Institute for the Study of Journalism.

Bucknell I. 2020. Going digital, not dying out: how universities are uniquely placed to teach digital journalism survival skills. Journalism Education. 9 (1), pp. 6-17.

Frost, A., Kynvin, J., Stuart, T. and Gallardo, S. F. 2022. *Communicating with Data Visualisation: A Practical Guide*. London: Sage.

Gye, H. 2017. SICK TERROR LIES Seven Manchester attack pictures circulating on social media that are FAKE. *The Sun.* [online]. 23 May. [accessed 29 July 2024] Available here: <https://www.thesun.co.uk/news/3631792/six-manchester-attack-pictures-circulating-on-social-media-that-are-fake-as-twitter-users-rush-to-share-information-in-wake-of-suicide-bombing/>

Innes, M. 2020. Techniques of disinformation: Constructing and communicating 'soft facts' after terrorism. *The British Journal of Sociology*. **71**(2), pp. 284–299.

Ireton, C. and Posseti, J. 2018. *Journalism, fake news & disinformation: handbook for journalism education and training.* UNESCO.

Karlova, N., and Fisher, K. E. 2013. A social diffusion model of misinformation and disinformation for understanding human information behaviour. *Information Research*. **18**(1).

Martyn, P. H. 2009. The Mojo in the Third Millenium: Is multimedia journalism affecting the news we see? *Journalism Practice*. **3**(2), pp. 196–215.

Newman, N. R, Fletcher, R., Robertson, C-T, Arguedas, A. R. and Nielsen, R. K. 2024. *Reuters Institute Digital News Report 2024.* Oxford: Reuters Institute for the Study of Journalism.

Petre, C. 2018. Engineering Consent. *Digital* Journalism. **6** (4), pp. 509–527.

Silverman, C. and Alexander, L. 2016. How Teens In The Balkans Are Duping Trump Supporters With Fake News [online]. 3 November 2016. *Buzzfeed* [accessed 18 July 2024] Available here: <https://www.buzzfeednews.com/article/craigsilverman/how-macedonia-became-a-global-hub-for-pro-trump-misinfo>

Thomson Foundation. 2021. *Remote control: Covid-19 and the smartphone.* [Online]. [Accessed 9 April 2024]. Available from: <https://www.thomsonfoundation.org/latest/remote-control-covid-19-and-the-smartphone/>

Wardle, C. and Derakshan, H. 2017. *Toward an interdisciplinary framework for research and policymaking.* Council of Europe.

Wardle, C. 2019. *First Draft's Essential Guide to Understanding Information Disorder.* First Draft.

Working safely online

Social media platforms are places where journalists can find and monitor a vast number of stories and contacts. You can spot and follow trends, gather different perspectives and keep abreast of what other journalists are doing. You can easily engage with your audience and immediately see their reactions to particular issues or stories. Having a personal presence on social media enables you to promote your work and to build your own profile. But for all the benefits and opportunities that digital platforms bring to journalists, there are also concerns. It is hard to quantify but there is increasing evidence of the abuse that journalists can face online.

In a survey of members of the NUJ (National Union of Journalists, UK) in 2020, just over half of the journalists who responded said they had experienced online abuse. Another recent survey, this time by the International Centre for Journalists (Posseti and Shabbir, 2022) spoke to journalists from around the world and found that nearly three quarters of respondents who identify as women had experienced online violence in the course of their work. Women journalists affected by other types of discrimination such as racism and homophobia were even more likely to be targeted, and with more severe impacts.

Although all the research shows that women and minority groups are more likely to face online abuse, anyone could be a target. There are numerous examples of journalists who have had to experience this, we share a few here as an illustration.

In her book *How to Stand up to a Dictator*, journalist Maria Ressa details the online (and offline) threats and personal abuse she received while trying to combat misinformation and corruption in the Philippines.

In 2022 Alec Belfield was sentenced to five and a half years, after being found guilty of stalking four people – including BBC radio presenter Jeremy Vine. Vine described being 'broken' by the experience and afraid of what might happen to him and his family, saying if the courts hadn't acted 'somebody would have died' (BBC Newsnight, 2022).

It's important for the digital journalist to be aware of the potential harm from working online and think about those risks and how to minimise them, ahead of posting. In this chapter, we'll discuss some of the dangers and the ways you can protect yourself and we'll signpost to sources of further advice and support.

What is online harm?

There are different types of online abuse and they can be difficult to define. As new technologies come and go, the threat can change. The World Economic Forum (2023) uses six broad categories to define types of online harm. Using these as a framework we can explore the kind of abuse that journalists are most at risk from, in terms of their profession.

- Threats to personal and community safety: This can include any kind of threat – physical, legal, reputation – to you or even to your family or friends.
- Harm to health and well-being: Exposure to traumatic content.
- Hate and discrimination: Hate speech is any kind of communication in speech, writing or behaviour that attacks or uses pejorative or discriminatory language with reference to a person or a group on the basis of their inherent/ protected characteristics – in other words, based on their religion, ethnicity, nationality, race, colour, ancestry, gender or other identity factor.
- Violation of dignity: Online bullying and harassment, for example personal abuse or trolling – negative comments about the way you look or behave. This could also include backlash – a sudden strong reaction to something you have said or posted. On social media this can become very large very quickly, often with groups of people making an organised effort to target you. Many of those taking part will not have seen the original comment or have any idea of the context. It could also include long-term harassment and online stalking – abuse directed to you over a long time. Intrusive or persistent messages could come across multiple platforms.
- Invasion of privacy: This could include revealing personal information about you, including doxing (searching for and spreading your personal details with malicious intent). This could be posting your home address on a chat site, or finding pictures of you that you might not want to share widely.
- Deception and manipulation: Including impersonating – creating a fake account that looks like it is you. This could be to use your professional credibility to spread false information, or to damage your reputation by posting things you

would not post. Hacking – where someone gains unauthorised access to your computer or accounts. They could for example hack your social media account and post something that would then look like it comes from you. There have also been examples of deepfakes where AI (Artificial Intelligence) is used to manipulate or create realistic images or videos that make it seem to you have said or done something you have not.

The most common of these is personal abuse, often from unknown or anonymous sources (UNESCO, 2020).

Often the behaviour is limited to the online sphere, but the online threat can transfer to the real world (for example online stalking becoming physical stalking, or threats or actions of physical or sexual violence).

The mental health impact on the victims of online abuse can be severe with many reporting anxiety and depression as a result. Washington Post technology reporter Taylor Lorenz has been open about the online abuse she receives online. In an interview with MSNBC (MSNBC, 8 April 2022) she talked about the impact, saying: 'I have severe PTSD from this. I contemplated suicide, and it got really bad.'

It can also impact on your professional work. An international report into online harassment of women journalists (Ferrier, 2018) found that 37 % avoided covering certain stories because of abuse, and 23 % had trouble establishing rapport with interviewees. Almost 30 % of respondents to the survey said threats and attacks had led to them considering leaving the profession.

Of course, everyone who uses social media platforms is at risk of online harm, but the profile and public perception of journalists can make them more of a target. A shocking 78 % of UK journalists (NUJ, 2020) agreed that abuse and harassment had become 'part of the job'.

This should not be the case, and measures are being called for, and taken by organisations and governments around the world to provide better support and safety for journalists.

In 2021 Dr Rebecca Whittington was appointed as online safety editor for the UK's largest news publisher, Reach plc. An industry-first, her role involves supporting not just the company's 3,000 journalists but also their readers. She's clear about the challenges facing journalists:

> Social media in so many ways is open for everyone so we get a vast array of representation and voices coming through from it, which actually, we wouldn't necessarily have had the opportunity to hear before. So, I think there's loads of positives about online. Sometimes it can be quite hard to see them because it can feel really scary, and it can feel intimidating ... But if you want to work in journalism, this is going to be your workspace. (Dr Rebecca Whittington)

In this chapter we'll outline some concerns and offer practical exercises that you can take to help protect yourself, so you can take advantage of the benefits of working online, while keeping yourself and others safe.

Your presence online

Think about what information you're willing to share about yourself. Connecting with the audience and showing personality can be powerful tools for a journalist and sharing real-life experiences can add to that. But it's always worth considering a worst-case scenario situation. Are you providing enough information for an online troll to use against you or put yourself or your family in danger?

Be aware of the *jigsaw effect* where over a number of posts you could give away enough small pieces of seemingly harmless information, that when put together create a much clearer picture. An example would be using a child's first name in one post, a picture of school in another, a birthday message in another and an anonymous photo in another.

An effective way to illustrate this would be to look at the content created by TikTokker Kristen Sotakoun (notkahnjunior) who has a series on 'consensual doxing' where people challenge her to find out information about themselves by searching online. Using their social media profiles as a starting point, Sotakoun is able to locate a vast amount of detail about her subjects, even those who believe they have kept information private.

Exercise 7.1: Explore how much personal information is readily available on social media

Try your hand at consensual doxing.

You could try this exercise with a friend or fellow student – make sure you both agree that you are willing to take part.

Look at each other's public social media accounts and use your investigative skills to see how much personal information you can find. Do you know how old they are? Their birthday? Where did they go to school? Where do they live? What are their hobbies? The names of friends or family? Clicking on linked friends or colleagues' accounts may give you even more information.

Share what you have found with each other. How accurate is it? Are you surprised? Are you comfortable with the amount of personal information you have shared online?

If you don't have someone else to work with, you can look across all your own accounts and see what you can find out.

Being transparent and sharing some personal details can be seen to build your authenticity and hence add to the credibility of your work (Hayes et al, 2007). However, it's up to you how much information about yourself you want to make public. Some people may be happy to share some relevant details of their private life, others will prefer to stay anonymous. The important thing is that you understand the implications. For example, a local news presenter may want to make it clear that she lives in the area that she reports on, but never mention her street name, or post pictures taken from her house.

Dr Whittington encourages journalists to write a physical list of what personal details you are willing to share online. They can then refer to the list every time you are thinking of posting.

It might be for example, if you have a young family, you might not want your children's names to be known by the general public – or for the general public to even know that you have a family. It might be that you go for coffee in a local coffee shop and you really enjoy it, but do you want the general public to know that that is where you go for your daily coffee?

Having this physical list gives you what Dr Whittington refers to as a 'stop point' – a moment to pause and think before pressing send, even during a breaking news scenario, which allows you to consider and double check what you are about to publish.

Exercise 7.2: Create a personal online safety plan

Take a piece of paper and draw lines across the centre horizontally and vertically, diving it into quarters. (You may prefer to do this on a computer rather than by hand – that's fine!)

This document will eventually be your *personal online safety plan*, which we will complete throughout the chapter.

In the first quadrant in the top left – hand corner, write a heading along the lines of 'Personal information I'm willing to share online'. Beneath that put a list of what information you would be willing to share about yourself with the public. You may also want to write down the things you absolutely never want to share. Refer to it regularly and check it before you post anything online.

You should try and create a stop point before every post you publish. Every social media post you send is a public reflection of you and perhaps the organisation that you work for. Do you want the general public to see that information? As Dr Whittington advises: 'That thing that you have written or sent a picture of… imagine that it could be got hold of by somebody that you don't want to get hold of it. Are you happy with that?'

You will want to carry out an audit of your own social media. Check that there is nothing there that you would not want the wider world to see. As we grow older our opinions or our behaviour may change. With fresh eyes, do you regret sharing a particular joke, or posting a particular photo? Many news stories have come about from journalists finding controversial posts that public figures have made on social media in the past. People who want to instigate conflict, hostility or harm online (commonly known as internet 'trolls') can do the same to you. It's worth turning your journalism skills on yourself and removing anything you think may cause problems.

It could be if you start working for a particular news organisation that you are asked to remove social media that does not align with their values – perhaps showing your political bias or reflecting negatively on your professionalism.

Another simple safeguard is to check your privacy settings. Your personal social media should be shared only with friends. That doesn't mean it's impossible for someone to find it – but it will make it much harder. Each social media platform has a different way of doing this, but it is usually found under 'Settings'. Familiarise yourself with what the settings mean and how to change them.

Many journalists will have separate personal and professional social media accounts. Bossio and Holton (2021) found this a common 'disconnection strategy' that journalists use to manage their relationship with social media:

> Journalists ... suggested that they chose not to post professional work on personal profiles or not share opinions on a professional profile; they engage in an ethics of disconnection because they extend a duty of care toward personal friend networks who might not find their work relevant to their interests and a professional duty of care to audiences following their work on social media and expect the expression of particular professional values

Having a separate personal account helps to keep your private information secure. It also means you have the option to mute, change or even delete a professional account while still being able to engage with family and friends.

If you are setting up a separate professional account, remember this will be the public-facing view that you present to the world. You will want to consider how much time and effort you need to put into maintaining it. Future contacts and contributors may well use it as a means of checking you out before deciding whether to share a story with you. Other journalists and potential employers may look at it as part of a recruitment process. If you leave this account unattended or incomplete it may have a detrimental effect on the way people see you.

Managing your relationship with social media

It can be hard to 'switch off' from social media. As a journalist you are naturally curious and want to keep abreast of what is happening. As a professional you want to see how the audience respond to your work – whether things are going well or going badly. But constant scrolling can take up so much of your time and be mentally exhausting. It can impact on your sleep and your overall physical health as well as on your mental health.

As tempting as it is to keep checking how many people have commented on your post or watched your digital short, it's important to take a break. Make rules for yourself about when you are going to access social media – and most importantly, when you are not. Make sure you take time away from your computer and your phone to balance your life. Check your phone and see how you can silence notifications from certain apps or at certain times. If you can, keep your work and personal phone numbers and email accounts separate so you can contact friends and family without having to check in on work at the same time.

Contacting people on social media

Be professional and honest: Think carefully about how you approach someone. Remember that the person you are contacting will also want to keep themselves safe, so introduce yourself properly and explain who you are working for and what you want in a way that makes them feel confident you are legitimate. Let them know how you found their contact details.

Put yourself in their place: Someone with experience of the media may be used to being asked questions, but a member of the general public might find it overwhelming, so guide them through the process. Explain why you are contacting them, why their input is important and how it can help to tell the story. Ask them how they are. Remember to treat everyone you reach out to as an individual not just a useful contact for a story. They may have an emotional response to what you are asking them.

Consider your message: Are you contacting them via a private message or via a public post (e.g., replying to a comment)? It is sometimes easier to contact someone via a private message, but the person you are contacting may feel that this is more intrusive. It may feel like a private message is safer but remember that these messages can still be screenshot or shared, so apply the same scrutiny to what you are doing as you would for a public post.

Don't post personal details: Don't share contact details that you are not prepared to share publicly. Use a professional email address and avoid posting personal details like phone numbers, location, etc.

Have a plan: If you're keen to get hold of someone it's tempting to send multiple messages via every platform but take a moment to consider how it might feel for that person to receive that many messages. Look at where they are most active and most likely to see and respond to you. Try that method first.

Be cautious: Have a healthy dose of scepticism about new contacts. Are they who they say they are, and are they telling you the truth? You could ask for verification (for example a work email address), check their social media, or search for them online.

Safeguarding yourself and others

Your safety is important, and it's vital that you consider how to keep yourself safe at all steps in working on your story – right from the initial planning stages.

Dr Whittington believes preparation and planning are an essential part of using social media safely:

> One of the key things that I do in my job is encouraging people to think about the risks of the story in terms of their personal safety and online and physical safety before they actually publish it – and encouraging the whole newsroom to think about that as well. (Dr Rebecca Whittington)

It's widespread practice in newsrooms around the world for journalists to complete a risk assessment before they go out and work on a story. This means considering potential problems, from minor hazards such as tripping over wires or being out in wet weather through to the life-threatening risks from taking part in extreme sports or reporting from a war zone. It's important to think about all the people who might be at risk during the activity, including reporters, crew members, contributors and passing members of the public – with special consideration given to anyone who may be vulnerable (e.g., have a disability, be very young or old, be pregnant). Once the potential risks from each activity have been identified, you can work out the actions you need to take to reduce or eliminate that risk (for example, using wireless equipment, packing raincoats or identifying shelter, using safety equipment or additional security).

This risk assessment process for physical newsgathering is now part of the everyday process in most newsrooms. As a digital journalist you can use the same process to consider the online harm that may be caused by working on a story. Examples may be the potential of backlash directed towards an individual reporter if you post an article on a controversial topic; potential identification of a vulnerable contributor; damage to reputation from posting inaccuracies. Taking the time to think about the 'worst-case scenario' ahead of publication can help you put protective measures in place. You may decide to amend details, to avoid tagging individuals, to turn off comments on social media or to identify sources of support in advance. In some cases you may even decide to avoid posting a particular story on certain platforms.

Remember to consider the impact on your contributors too. Particularly consider young, vulnerable contacts. Explain to your interviewees when their story will be posted online and what reaction could be. You could suggest that they check their own social media accounts and privacy settings.

One of the simplest but most effective things you can do to protect yourself is to make sure someone else checks your work before you publish it. Having this 'second pair of eyes' looking over your work is essential.

In traditional media it's virtually impossible for a journalist's work to be published or broadcast without a sub-editor, editor or more senior journalist's approval. The immediacy of social media means it's easier to post yourself and tempting to get a story or comment out as quickly as possible. However, taking time to draft your post first, check for inaccuracies or for anything that might be misinterpreted – and then getting someone else to do the same – will drastically reduce the chances of you making a mistake, and thus reduce the fuel that trolls can use against you.

Exercise 7.3: Make a risk assessment for a story you want to work on

Think of a story you would like to work on and consider the risks and then any measures you could put in place to minimise those risks.

You may find it helpful to use a risk assessment template; you can find some of these online.

Protecting your data

Keeping yourself and others safe from harm also means making sure that you take steps to protect your data. You should follow data protection laws about the storage of personal data at all times. These may be different around the world so check your local law and guidance for specifics. Here we will concentrate on practical steps a digital journalist can and should take to avoid common forms of online harm.

- Second factor identification: Set up your mobile phone, laptop, tablets to require second factor identification. Most systems will have an initial password plus perhaps fingerprint recognition, face ID or a single use pass code sent to another device. Whilst this can be frustrating in taking up time and needing extra steps to get into your own work, it will make it much harder for anyone to access your data should they get hold of your equipment.

- Back up your data: If everything is on your laptop and then your laptop breaks down or is damaged or lost, you may have no way of recovering your work. If you are backing up onto a physical drive rather than a cloud-based server, make sure you keep that drive somewhere safe and password protected.
- Check your passwords: Are they secure and do you change them regularly? Don't use the same password on different accounts. Don't write your passwords down. There are secure password managers that you can use to keep track of and store your passwords if you are concerned about this.
- Share data safely: Don't pass on private or confidential information to other people, or save it anywhere insecure, or put it into an AI on online tool that you can't guarantee is protected.
- Be wary of scams: Never click on links in emails from people you don't know. Don't install software that you are not sure where it comes from.

Dealing with abuse

If you do find yourself a victim of something that upsets you online, there are a few steps you can take.

Firstly, talk to someone you trust. It could be someone at work or someone in your personal life. You don't have to go through this alone and sharing your concerns with someone will help you get some perspective over how big the problem is. We are all inclined sometimes to ignore a hundred positive comments and focus on the one voice that is negative. Talking to another person can help you realise that. It is also helpful to concentrate on the response of people who know and support you, and whose opinions you respect and value, rather than the messages from strangers online.

When you are in a work environment there may well be support provided through your workplace, make sure you are aware of what is available.

Exercise 7.4: Make a list of people you trust who can support you

Think about the people you could talk to for support if you were worried about online messages. Include contacts from both your professional and personal life. These should be people who you trust and whose opinions you respect.

Write their names in the top right-hand corner of your personal online safety plan under the heading 'People I can talk to'. You may also want to include the reasons why you have chosen that person. If you are feeling worried or less confident you can always look at this list and it will remind you whose advice you value the most.

Responding

You have the right to respond to any comments made to or about you online. You will want to exercise caution and think carefully before doing so.

Online trolls, like physical bullies, are often trying to provoke an emotional reaction to what they are doing. These trolls rarely engage in a conversation, change their opinions, or acknowledge or apologise for hurt caused – so any response you make is unlikely to make any difference to their thinking. Often the most effective technique to deal with it is to refuse to engage and prevent the situation from escalating. There may be occasions where you feel you must, or want to, respond to a message or comment.

Unlike abuse, which is often personal rather than commenting on your work, criticism can sometime be valid and relevant and may require some kind of response or acknowledgement. There may be times when the point someone makes is true (even if it isn't always pointed out in the kindest manner) and you feel you need to acknowledge or even apologise for a mistake. Perhaps you feel a comment online is so wrong that you need to correct it. Or you want to publicly call out bad behaviour. If you do decide to respond, don't react in the heat of the moment. Talk it over with someone first. Allow the emotion out of the argument and respond in a professional manner. Draft a response and take some time to check it and think about how it will be received before posting.

Things move quickly in the digital world, and while a post or message may initially attract a lot of attention, it is often far down people's screens by the next day and soon forgotten. It may help to take some time away from social media and remove yourself from the temptation to respond and check your messages. You may find that the issue has quickly died down.

If you receive a particularly hateful comment or someone is persistently messaging you, you may want to consider blocking that individual. This will prevent them from being able to see your content or from contacting you. Different platforms have different ways to do this, so check that you know how it works for the platforms that you use. In extreme cases you may find that taking a break from social media, or one particular platform on a longer-term basis is preferable. Again, talk this over with someone and seek advice before you make your decision.

Reporting abuse

If you believe that you, or someone else, is in immediate danger, you should always call the police. If you are working for someone, you should let your editor or line manager know as soon as you receive any kind of abuse. If you are a student, let your tutor know what has happened. They will be able to offer support and guidance.

Each platform has a different way to allow users to report harmful content. For example, on Facebook you can click on the three dots at the top right of any post and choose 'Report Post' and then select from a list of concerns to alert the platform to a problem. From TikTok you would use the arrow to share a post and then select the 'Report' option. Reporting content not only makes sure the platforms know, but also may help protect other people.

Laws in different countries vary, but if you are a victim of persistent or extreme abuse or threats it's likely that a crime is being committed and you should officially report what is happening to the police. You will need to document what is happening to you. Some trolls will make content private or even delete content they know will get them into trouble – so taking screenshots will help. This means you can build up a case of evidence for if further action needs to be taken.

Don't feel you have to go through this alone. Make sure you have support. This may come from a colleague, friend or family member. If you are a member of a union, they will be able to offer advice. You can get further advice from a number of agencies and organisations, depending on where you live. We've included some useful contacts at the end of this chapter.

Exercise 7.5: Think about how you would respond to online abuse

Imagine that you have been working hard on a news story for a few days. Your editor has checked it and approved for publication online or on social media. It's had a lot of engagement, most of it positive – but you've also attracted the attention of online trolls.

Here are some examples of the kind of online abuse you could receive in response. How would it feel for you to read these comments? What would you do? Would you respond? If so, what would you say?

Commenting on your physical appearance/clothing
Questioning a particular fact in your story, and telling you that have got the data wrong
Accusing you of biased reporting
Saying you are stupid and bad at your job because you made a spelling mistake
Making assumptions about your life that aren't true (e.g., 'you're too wealthy to understand this issue')
Claiming you are incapable in some way due to your gender/age/race
Calling you racist/sexist because of the way you phrased something in your story
Suggesting people should go to your place of work and throw paint at you in protest over something you have said

You can think about this on your own, but it may help to talk it through in a group, so you can discuss it from different perspectives.

Dealing with trauma

As a news journalist you may find yourself covering stories where a traumatic or violent event has occurred. You may not be the direct witness or victim of the event, but you may be on the scene of the aftermath, interview people who have been affected, come face to face with perpetrators or hear details in a court case. In the rush to cover a news story, many journalists find they need to push these feelings aside in order to do their job. But the emotional impact is still there.

Journalism is one of the professions that puts individuals at risk of vicarious trauma, or secondary traumatic stress. That is indirect trauma received by being exposed to difficult or disturbing situations second hand. As a digital journalist, perhaps spending a lot of your time working at a computer screen, it can sometimes be harder to identify the potential for trauma. If you are attending the scene of a crime, or physically interviewing a victim, you are easily aware of the impact it is having on you.

But if you are scrolling through social media you can come across difficult content unprepared. Or perhaps you are looking at images from a breaking story without full processing their impact, messaging people who are telling their story for the first time rather than talking to them in person or repeatedly watching difficult videos to for verification or to look for details.

If you do see disturbing content online, report it. You can do this via the platforms, internet safety organisations, or the police. It can help you to know that you have done something to help protect other people from seeing the same thing. It can also be beneficial to talk about it to someone, rather than keeping it to yourself.

The effects of vicarious trauma can manifest in different ways for different people so look out for signs of changes to your thoughts (intrusive thoughts, guilt, hopelessness, fear, numbness), behaviour (mood swings, withdrawal from social situations, reliance on escape mechanisms such as alcohol) or physical health (headaches, stomach problems, fatigue). Looking out for these changes and addressing them if they occur can help you to keep yourself safe.

It's normal to be affected by some of the stories you will come across in your career. Some stories will impact you more than others. This can be different for everyone. It's not always a big and obvious story that has a negative impact on your mental health. It can be a build-up over time, a delayed reaction, a smaller but more personal incident. You may know in advance that a particular subject might trigger difficult feelings, or the strength of your reaction may surprise you. It's OK to acknowledge that you are finding something difficult and to ask for support. Even the most experienced journalists are affected by these issues. Talk to your tutor, line manager or editor. There are resources for support at the end of this chapter.

One proactive thing you can do is prioritise looking after your own mental health. Knowing how to care for yourself helps build resilience so you are more aware and able to cope if things do get difficult. It's important to practice self-care as part of your normal routine.

- Take breaks: Spend some time away from your work and do something that makes you feel good. Go for a walk, listen to music, relax in a hot bath, spend time on your hobby ...
- Look after your physical health: Eat a balanced and nutritious diet, stay hydrated, exercise. Maintaining your physical health will help your overall well-being.
- Sleep well: We all need rest, and a good night's sleep can make the world of difference. Stick to a routine. Turn off your screens before bed, avoid caffeine, pop on an eye mask ... whatever helps you to wind down and relax.
- Stay connected: Talking to your support network of colleagues, friends or family will help you feel less isolated.

Exercise 7.6: Create a self-care plan to bolster your mental health

Draw up your own self-care plan, using the list above as a guide.

Everybody is different, and we all have our own unique things that make us feel positive, stronger, rested and able to cope. Think of the things that work for you. It could be going to the gym, baking a homemade lasagne, watching your favourite film, talking to your best friend, lighting a scented candle, hugging your dog, sitting on a bench in your local park ...

Make a list of the top ten things that work as self-care for you and add it to the bottom left-hand corner of your *personal online safety plan*, under the heading 'My self-care actions'. Look at this often to remember to take time out and do the things that make you feel good.

How can you prepare and support your audience?

You should consider harm and offence in your digital journalism with the same editorial rigour that you would for more traditional forms of media. While you may have a target defined audience, you have no way of knowing who will consume your content and so you must assume it can be accessed by anyone – that includes young and vulnerable people.

If you know you are showing potentially upsetting images, or producing content about a sensitive subject, you may want to include a trigger warning. This is a message to be given before your content is viewed – in the text of the introductory post, or as a caption at the start of a video – alerting viewers to distressing content.

You may also want to consider how you can offer support to anyone who might be affected by the issues you cover. You could do this within your content by including expert advice, or by signposting your viewer to places where they can get help and information (such as relevant charity websites, support groups, etc.).

Support and resources

Here are details for some organisations that offer support and resources for journalists about online safety.

- Headlines Network: Promoting mental health in the media – <https://headlines-network.com/>
- Thomson Reuters Foundation: Practical and legal tools to protect the safety of journalists – <https://safetyofjournalists.trust.org/>
- PEN America: Online harassment field manual – <https://onlineharassmentfieldmanual.pen.org/>
- Dart Centre for Journalism and Trauma, Columbia Journalism School – <https://dartcenter.org/>
- NUJ Journalism Safety Toolkit – <https://www.nuj.org.uk/advice/journalists-safety-toolkit.html>
- UNESCO Safety of Journalists – <https://www.unesco.org/en/safety-journalists>
- Reporters Without Borders: RSF Resource for Journalists' Safety – <https://safety.rsf.org/>
- International Womens Media Foundation: Journalist and Newsroom Safey – <https://www.iwmf.org/safety/>
- First Draft News – <https://firstdraftnews.org/>

This is not a complete list of the support out there. You may want to look for organisations that are based where you live, or that concentrate on your particular field of journalism, or that support people from a similar community to you.

If you have an employer, are at university or are a member of a union make sure you are aware of any support that is available to you through those channels.

Exercise 7.7: Complete your personal safety plan with a list of helpful resources

Look at some of the websites outlined above, and search for additional sources of support.

When you find a resource that is useful to you, add it to the final quadrant in the bottom right-hand corner of your *personal online safety plan*.

Keep your completed online safety plan somewhere that is easily accessible. Refer to it regularly and update it when necessary. This plan will remind you of ways you can keep yourself safe and point you towards help if you do need it.

Table 7.1: Example template for a personal online safety plan

My Personal Online Safety Plan	
Personal details I'm willing to share	People I can talk to
My self-care actions	Support

References and recommended reading

BBC Newsnight (2022), Jeremy Vine: My jailed stalker Alex Belfield says he'll be back, *BBC News* [online] 21 September 2022 [accessed 16 July 2024] Available from: <https://www.bbc.co.uk/news/uk-62977842>

Bossio, D., and Holton, A. E. 2021. Burning out and turning off: Journalists' disconnection strategies on social media. *Journalism*. **22**(10), pp. 2475–2492.

Ferrier, M. 2018. *Attacks and Harassment: The Impact on Female Journalists and Their Reporting*, International Media Women's Foundation and Trollbusters.

Grant, J. I., Hildreth, A., Dobrygowski, D. and Bantourakis, D. 2023. *Toolkit for Digital Safety Design Interventions and Innovations: Typology of Online Harms*. World Economic Forum

Grossman, B. 2021. *MLA Safety Guide: Combatting online harassment and abuse – a legal guide for journalists in England and Wales*. Media Lawyers Association

Hayes, A. S., Singer, J. B., and Ceppos, J. 2007. Shifting Roles, Enduring Values: The Credible Journalist in a Digital Age. *Journal of Mass Media* Ethics. 22(4), pp. 262–279.

Kean, J., and Maclure, A. 2021. 'A modern-day equivalent of the Wild West': preparing journalism students to be safe online. *Journal of the Association for Journalism Education*. **10**(1), pp. 69–82.

Meet the Press Daily, 8 April 2022, MSNBC [online]. Available at: YouTube. [accessed 16 July 2024].

Posetti, J. and Shabbir, N. 2022. *The Chilling: A global study of online violence against women journalists*. International Centre for Journalists

Ressa, M. 2022. *How to Stand Up to a Dictator*. London: WH Allen.

Storm, H., Crowley, J and Berrie, A. 2023. Vicarious trauma: A guide for journalists and newsrooms to recognise vicarious trauma and mitigate against it. *Headlines Network and Mind*. [Online] [Accessed 1 August 2024] Available from: <https://img1.wsimg.com/blobby/go/6ca5410e-ac1d-4642-b85f-b717e6a71453/downloads/Vicarious%20Trauma%20Guidelines%20 2022.pdf?ver=1676045989580>

Wolfe, C. 2019. Online Trolls, Journalism and the Freedom of Speech: Are the Bullies Taking Over? *Ethical Space: The International Journal of Communication Ethics*. **16**(1), pp. 11–21.

Whittington, R. 2023. *Online harms against women working in journalism and media – Research report*. Women in Journalism

Future proofing

The aim of this book is to help you become a digital journalist and one of the main challenges is to do that in an environment where many of the systems that you'll use to make and share your stories, and the roles that you'll play as a reporter, do not yet exist. How do you prepare for life as a digital journalist not just now, but into the future? This chapter considers approaches to your work that will help you to stay on top of your brief no matter what changes computing systems may bring about.

We have already dealt with one key component of future proofing yourself as a journalist in the twenty-first century. In Chapter 6, *Journalism skills for the digital age*, we explained how the fundamentals of the profession endure no matter what production and news access trends come our way. The core skills (nose for a story, verifying information, interviewing, producing video, writing, etc.) remain the same. In fact, they seem to be particularly highly valued because they transcend medium, platform or format, whilst many other aspects of digital production can be ephemeral. As a BBC Yorkshire digital editor put it: 'as we move forward, I think we need to equip people to do the old things better'. Bearing this in mind, as students of journalism, we suggest that you don't get discouraged if your tutors set you exercises and reporting tasks that are not bound to online work. The skills and understanding you'll gain are essential to becoming a digital journalist.

This chapter highlights three further steps you can take, informed by industry experts and best practice, to make sure you're equipped to cope with whatever developments journalism will throw your way. We start by suggesting that one way of being certain that you'll not get left behind is by playing an active role at the forefront of developing new ways of telling stories using digital media. We'll focus on a method of innovating that is seen in industry that you can employ here in completing an exercise in new digital production. We'll also discuss some of the 'soft skills' that industry experts have deemed essential to thriving as a digital journalist. Finally, although it's never easy to predict what will impact on journalism next, artificial intelligence is already playing a large part in how journalists work and what they produce, so you'll also find out more about its influence on the newsroom.

Innovation: New digital formats

There's no better way of keeping up with the pace of change than being a part of it. Rottwilm wrote that journalism teachers in higher education should be active in meeting the challenges presented by digital media: 'Academics, so often external spectators, need to be offered a seat at the table in order to provide the insight and experience that comes from studying these issues in a deeper context and over a longer period' (2014, p. 20). We have a better idea. Students, with the support of their tutors, can take the initiative and become innovators, rather than waiting for an invite.

The task is to not only get on top of learning digital journalism, but to stay ahead by being a part of its evolution. Universities are already well-placed to help you cope with the pace and relentlessness of technologically driven change. We should also support you in participating in future developments, to understand what is happening from the inside, rather than wholly relying on what research might say, based on temporary relationships with news organisations. An appetite for exploration, innovation and continuous study, is shared by universities and digital journalism producers, so this seems like a solid basis for having a go at taking a front seat and seeing what new ways of reporting you can create yourselves.

Here's one approach you can take to innovation, focused on using social media to produce and disseminate stories. We want you to find a way of telling a story natively on a social media platform in any way you see fit. There's nothing specifically that we want you to do but what we want you to do is have a go, see how it works and learn from the experience.

The method we want you to use follows a pattern of adaptation seen in digital journalism, as news organisations fuse their traditional strengths with new production methods afforded by digital media. Look at these examples:

- Live blogs take a pre-existing form of journalism (rolling news coverage) and synergise it with web production:
  Rolling news + web production = live blogs

- The Financial Times has had a lot of success on Instagram with what it calls 'social cards'. As we've seen earlier in the book, they produce an infographic that visualises a story based on data and write a couple of paragraphs of accompanying text:
  Infographic + Instagram post = social card
- Similarly, the New York Times, harnesses its tradition of using visually impactful images to 'make complex stories more digestible and accessible' (Tameez, 2021).
  Photojournalism + Instagram carousel = photo story
- Video explainers also fall into this theory.
  Public service journalism + social video = video explainer

We can see a successful formula here that you can apply in creating a new approach to reporting a story:

Established journalistic form + Digital media = New story format

Of course, it's not the case that every new creation in digital journalism has been brought about in this way or conforms to this model, but the frequency with which it occurs gives us confidence that this approach will serve you well. Whatever the result, the most important thing is what you learn from the experience, so put perfectionism to one side and have a go at the following exercise.

Exercise 8.1: Create a report using a social media news format that you have developed yourself

Step 1: Create and publish a report natively on a social media platform. By natively, we mean tell the whole of the story within the media that you post, rather than leading the audience away from the platform by, for example, linking to an online article. You have lots of options and we want you to take risks and innovate. Have a go, see how it works and learn from the experience.

The formula to use is: Established journalistic form + Digital media = New story format

You can choose any subject and any type of story. It could be an issue, incident, event, quirky phenomenon, inspiring personal story, etc. In other words, the full range of stories you've encountered in your journalism studies. You can also cover any topic you want, however serious or soft.

Although you'll create different types of reports, the essential criteria are the same:

- Make sure your story choice, storytelling and chosen platform are aligned. A picture-driven piece on an emerging art form will work on Instagram (see Chapter 5).

- Find a story that will interest the platform's audience and immediately capture their attention in the crowded social media marketplace.
- The normative standards of journalism apply get your facts right using well-sourced information, provide insight from relevant contributors, tell the story clearly using direct language and by linking each element of the narrative before bringing it to a conclusion.
- Use a tone of voice and presentation style that fits both the story and social media you're using.
- Most of all, bring creativity and a willingness to experiment. The best way to express your understanding of digital journalism is to create original work.

Step 2: As with any other form of experiential learning, maximise its value by reflecting on it afterwards. Alongside your own thoughts, get feedback from a tutor and peers. Use all of that in your analysis of what went well, what didn't, what challenges you faced and how you addressed them. Then create action points; two or three ideas of things that you can practise or research that will improve your work in future. For example, if the feedback says your story takes time to get going, create a new opening that has more of a hook. If you're not happy with your images, make time to practise your shooting and/or graphic creation skills.

Look up the Driscoll Model for reflection to help guide your considerations. You could present your findings in a short essay but, given the nature of our subject, a video explainer-style presentation of your findings would work well.

Coping with constant change

Whether it's the software we use to research stories, the devices deployed to record and produce them, or the platforms and formats preferred for telling them, one thing we can be certain of is that few things stay the same for long with digital journalism. As you have seen in this book, even a relative staple such as the social short is constantly evolving, with changes to the ideal duration, filming style, presentation and graphical content. This creates headaches in newsrooms, as highlighted without prompt by editors from a range of UK news providers (JPI Media, ITV News, Financial Times and BBC Yorkshire), who see keeping up with the pace of change as a major challenge to them as digital journalism leaders. They explained that computational advances and the power of third-party platforms were the main drivers behind this constant state of flux.

The impact this has on news work is captured by Ryan Daniels at CBS Sports, who says that strategies are 'led by the social overlords'. He cites how when TikTok came along and 'blew everyone out of the water', rivals such as YouTube wanted to follow suit and created YouTube Shorts, encouraging their content providers (what we used to call journalists!) to focus their efforts on these impactful videos of no more than a minute.

The pressure is on to fall into line, as the BBC's Jacob Tomlinson confirms: 'you're at the discretion of these third-party platforms so, ultimately, you can make the best content in the world, but if it's not going by their rules, it's not going to go anywhere'. At Channel 4, Georgia Carroll reports that they have strategists specifically employed to keep an eye on platforms to see how they are changing, such is the dependence on them as the route to finding your audience.

A digital editor at ITV News says the resulting velocity of evolution within digital journalism is unprecedented in the industry:

> It's a blessing and a curse really working in digital in that it's forever changing. With TV formats, if you'd have watched a bulletin from 50 years ago to now you wouldn't see a massive difference. Maybe one's black and white and one's colour but you wouldn't see a massive shift. Whereas with digital in the last couple of years, we've shifted considerably, and I think we continue to and every time we bring something new in or try a new format, we know at the back of our mind we'll be changing this again in a few months' time. So, it's great and exciting in that way and we have to keep developing and moving to try and stay ahead of the curve but sometimes it's hard to keep up. (Bucknell, 2020, p. 14)

Regional newspaper group JPI Media gives us an insight into how the pressure is experienced on the newsroom floor:

> We have emails going out constantly to our staff saying this has been changed now, would you do it this way? This is changing from next week, etc. Constant adapting. It can be quite tiring in a way, constantly changing to the way social media platforms are changing. (Bucknell, 2020, p. 15)

Newsroom managers, such as our digital editors, not only have to cope with the stress that they feel as a consequence of ever-changing strategies, skills and practices, but have to bring their teams along with them. This is where you come into it. The head of digital news at the FT suggests that one approach that seems to work when coping with constant change is to recruit new reporters who bring with them the ability to adapt: 'we are asking [of candidates] that there be evidence of awareness and a willingness to be able to be a quick study or a good student of whatever a newsroom's need is'.

The desirability of being a good learner and somebody who can handle or even thrive on change suggests that graduates of journalism schools in universities are well-placed to succeed as digital journalists. You're, amongst other things, being taught how to teach yourselves and find solutions independently. You're learning the mental agility to be open and responsive to new ideas and innovation.

What the university sector does as a matter of course, seems perfectly in sync with the key underlying characteristic of a successful digital journalist.

Georgia describes a mindset around her studies of the subject that helped her at the University of Leeds and into her role at Channel 4: 'I kind of accepted when I started the course that the job I would do didn't exist yet, just because media changes that often'. Be like Georgia and embrace how the stream of challenges you'll face as a journalism scholar not only allow you to succeed on your programmes but actively set you up with a key soft skill required for working as a digital media professional.

Don't sweat the small(er) stuff

Further solutions to the problem of how journalism programmes cover a vast and volatile breadth of digital activity can be distilled from what our digital editors had to say. Many emphasised the desirability, in some cases, of raising your awareness of new practices, formats and genres, rather than filling your time and brains with specific processes, software and skills that you may never use or could acquire on the job. A UK editor for vice.com used learning the Snapchat content management system (CMS) as an example. They explained that, from their point of view, in a couple of years' time, their publication may no longer be on Snapchat. Even if it's, the CMS is likely to have changed by then, so specific knowledge of menus, tools and other options will be out of date. But even if vice.com is still on Snapchat and the CMS has not changed significantly, it's the kind of system they believe a bright graduate will be able to learn in a few hours, so long as they have a working knowledge of how a CMS functions, for example, by using WordPress for online production. What they cannot teach in one morning are the aforementioned core journalism skills, a knowledge of native content creation and complex production tasks such as filming and editing.

Virtual reality (VR) and 360-degree production are examples of practices that many interviewees (BBC Yorkshire, ITV News, vice.com and FT) expected graduate employees to have an 'awareness' of due to their emerging role in digital journalism, without requiring them to have expertise or experience of using the technology for production. Some of our news organisations had experimented with VR and 360-degree storytelling, but none of them have taken it forward as an ongoing format. JPI Media explained that it's, 'not really in the core of the business at the moment.' It's a similar story at BBC Yorkshire, whose digital editor says not every form of digital media is going to work for journalism:

> I think, when there is a new tool out such as 360-degrees we rush to all use it, and then what you see after two months is a real recession in the use of it and then it comes back

> intermittently. And the reason for that is we like new things as journalists, and we can use them. However what we don't do is identify what advantage they give us or what's good for the audience. (Bucknell, 2020, p. 13)

ITV News was a little more effusive, saying that they have 'done a few 360 videos, as in crowd shots, where it's lent itself to that', and might find more uses for it and VR in the future. One way in which you could practically raise your awareness of augmented reality in journalism is to imagine how you might report on a story using the technology. Have a go at *exercise 8.2*.

Exercise 8.2: Plan a news report for new technology.

Use the principle outlined earlier in the chapter to match an established journalism reporting form with immersive technology. How would you tell your story for someone who is using a VR headset? How could you tell your story for someone who is listening on a voice-activated smart speaker at home? Could a hologram help you bring your story to life?

Comments made about learning Search Engine Optimisation (SEO) techniques are similarly cautious:

> There is an expectation of at least basic understanding of how to tag your articles, relevant keywords in the title or the subtitle, but yeah I'd say that's kind of a minor concern, that's more handled by, at least in our company, by social media department and the marketing department rather than the journalists themselves. (vice.com)

BBC Yorkshire also did not see SEO as a core skill for digital journalists: 'They need to know that SEO exists and a basic ten-point SEO list, and they might spend five hours doing it but I'm not convinced they need to know the bells and whistles of SEO to be honest with you'.

That thought is echoed by the FT whose digital editor does not expect media graduates to be able to do everything: 'we are asking [for] an awareness, if nothing else, of not how to produce everything, but of how to translate journalism into being relevant to whatever publications, platforms are necessary'. This is the key point. It speaks directly to the approach put forward in this book, that one of the secrets to learning digital journalism is that you do not need to cover every facet of practice in depth. So long as you have a wide-ranging knowledge of what is happening in the industry, learn the principles of digital journalism, acquire core skills and have a willingness and capacity to learn, you'll be fine.

Resilience

Sigh. It's a message you'll have received from old people throughout your lives and you may well feel that you've heard it often enough, 'thank you'. But a panel of recent Journalism graduates from the University of Leeds returned to the need to be resilient as key to coping with life as a new entrant to digital journalism. The issue of coping with constant change is one example of where this trait comes to the fore, as you'll be pushed and pulled in different directions according to forces that you cannot control.

Jacob from the BBC says resilience is also important when it comes to seeing how well your stories have performed according to the analytics. He talks about 'having a thick skin' and 'learning not to become too attached to the content you're making'. He gives an example of the sense of injustice you might feel:

> The chances are you could post content and it absolutely cans. I think I learnt that the hard way when I started. I would get so invested in my piece, think really outside the box how to do it and make it look classy in the edit, really smooth. I'd get it in and it'd get 10,000 views and I'd be like, 'oh, great, cheers'. (Jacob Tomlinson)

By comparison, he says, they might post some user-generated content (UGC) that 'someone filmed on a potato', and it might get five million views. We think he's exaggerating here, but you sense his frustration. Jacob advises that you must accept the give and take, be 'flexible and understand that another week the same piece you lovingly produced will do really well'. He also advises ending on a positive by reflecting on your work and thinking about what you could do differently and better next time. Keep going. Be resilient!

Another experience you're likely to have where the resilience jacket will need to be worn is in dealing with commissioning editors who are, shall we say, senior to you not only in rank. At times, they simply won't get it. While you're social natives, many of them honed their craft before social media was even a thing. Jack Gilmartin from the digital team at Channel 4 explains that the people who control the editorial purse strings 'have worked in a linear landscape for the last 20–30 years and the hardest thing is trying to like sell our ideas to them'. But he hasn't given up and sees it as his role to educate colleagues from legacy media backgrounds on the ways of digital communication. The proof of the pudding is also in the eating and Jack has experience of making a breakthrough with a commissioner, showing them what can be achieved with an innovative approach and therefore gaining their trust.

AI in the newsroom

Emergence of AI

The launch of ChatGPT in November 2022 by a little-known tech firm called OpenAI, brought generative artificial intelligence (AI) into the mainstream.

Trained on vast amounts of text and data (known as a Large Language Model or LMM) the system was able to answer questions and carry out tasks in the form of a chatbot the like of which we'd never seen before. It had an instant and profound impact on business, politics, education, personal lives and, yes, journalism. As a technology that can both source information and output it in a form desired by the user, it has 'obvious and profound potential impact for journalism [and] is now the urgent focus of the senior leadership teams in almost every newsroom' (Caswell, 2023).

In fact, AI is already contributing to every part of the reporting process, from finding stories, through research, writing, generating media and distribution. Uptake varies widely throughout the world but is extensive and growing. A report by the London School of Economics, drawing on the experiences of 105 news groups from forty-six countries, found that 75 % of respondents use AI for at least one part of the journalistic process (Beckett and Yaseen, 2023, p. 6). The amount of research and writing on the impact of AI on newsrooms is already extensive. These studies reflect a growing grasp of key issues (global inequalities, intrinsic biases, threats to existing jobs, etc.) and opportunities (speeds up repetitive processes, makes time for more investigative journalism, creates new jobs, etc.).

But what has struck us most in our review of the available literature is how the experience of adapting to and adopting AI is similar to previous experiences of journalism changing because of advances in computing. The same themes emerge and a well-established narrative continues. As with digital journalism generally, AI evokes a mixture of excitement and horror; brave new worlds and dystopias. It's not just the tangible practical benefits and deficits of the technology that impacts how it's viewed. How it makes people feel matters too and AI as an aid to journalism is met with some scepticism, such as worries about the threat to journalistic values, the lack of trust in third-party providers and a general fear of change.

Another familiar theme is raised by Caswell, who makes the point that rather than using generative AI (tools that create new content) to support existing routines within established news dissemination models, organisations should be open to new ways of working, creating new products that make the most of the specific benefits of AI. He exemplifies his point with reference to the Artefact news aggregation app, that harnessed AI to make personalised journalism recommendations to users, suggesting news outlets create a similar experience on their own platforms.

The journey from working in old ways with new tools, to finding new ways that are in step with how people use the new tools, is familiar. It's reminiscent of how news organisations were slow to adapt to the audience experience of social media. At the outset, they regularly copied and pasted content from legacy media stories into social media posts. For example, the top line from the online news report would become the post on Twitter or the video made for a TV news bulletin would be uploaded straight to Facebook with no edits.

Another familiar part of the digital journalism transition process that applies to working with AI, is that the fundamental skills of journalism are predicted to endure:

> The ability to maintain a keen awareness of the deep informational needs of an audience or society, identify stories that meet those deep needs, verify and contextualise those stories, and then communicate them to audiences in clear and engaging forms – will probably remain the foundation of journalism (Caswell 2023).

But with these foundations still in place, how we build on top of them will change in step with the software and our understanding of how to use it best. This leads us back to another principle of digital journalism practice established earlier in the chapter; a key quality of a news employee in an AI-driven future, is the capacity to thrive on change, as 'it is likely that news organisations will place a higher priority on adaptability and constant entrepreneurial innovation' (Caswell, 2023).

This is good news for you, with the implication being that as journalism graduates, arriving with knowledge of AI tools and the expectation of having to be flexible around work routines, you're at an advantage. Another way in which you have an edge, touched upon previously in our considerations of preparing to become a digital journalist, is that you have grown up amongst rapid advances in computing and can follow that flow more readily than many of the people who'll be looking to hire you: 'Generation Z's positionality toward technology can help to explain why being AI-literate is not an obstacle for this generation but a rich opportunity for development' (Salgado, 2022, p. 94). Cue you becoming active learners, innovators and practitioners with artificial intelligence.

What we have seen in this reflection on the early stages of AI reporting is that when it comes to technology events, journalism reacts in established and predictable ways. It's a similar process to Diffusion of Innovation Theory, that considers how new technologies spread through societies, with innovators and early adopters leading the way, followed by the majority. AI is the latest chapter in the ongoing story of how advances in computational power impact on journalism, with the same patterns of resistance, acceptance and adoption.

Which leads us nicely to a more detailed look at how AI is used in newsrooms and how you can bring it into your own practice. In the remainder of this chapter, as Caswell puts it, you'll 'navigate a path that takes advantage of near-term opportunities for efficiency, medium-term opportunities for new products and services for audiences, and longer-term opportunities for re-imagining what news can become in a transformed information ecosystem' (2023). Our intention is not to cover every aspect of AI in detail – there are other places you'll find this information if you want it (see the *further reading* suggestions) – but rather to give an overview that will help inform your understanding of the potential benefits and pitfalls it offers you as a journalist.

AI for finding news

If you've ever searched for something online, you've used an AI tool – a technological solution that replicates a job previously undertaken by a human. In this case, rather than heading down to a library to use a card indexing system and microfiche (you can search online to see what that means) as we did in a pre-computer age, the search engine can automatically sort through millions of pieces of information and find what's relevant to you. It takes a tiny fraction of the time it would take to do the job manually, and in most cases the outcome is actually more successful. It allows you – the human element of this task – to spend less time hunting for sources and more time thinking about the subject you're researching. In an ideal world, this is what technological advance is all about, making processes more efficient so that we can concentrate on other areas.

Newsrooms already make use of AI to help with these processes. Computers are much more efficient at large data analysis than humans. AI can be used to sift through large amounts of information and can spot where there could be trends or anomalies. These trends, differences, outliers are often where the news story lies.

An example would be using AI for event detection. If you have a computer system constantly monitoring social media output, it can notice where there is unusual activity. This means it could instantly spot where, for example, there is a sudden spike in people posting updates from one specific geographical area – alerting the newsroom to a potentially breaking story that may be worth looking into.

AI for distributing news

Another area where AI is being successfully used in newsrooms is in the processing and distribution of stories. Tools can be used to automate the identification of tags and key words in articles, speeding up and making more efficient the process of Search Engine Optimisation. AI can also be used to analyse audience data, spotting trends in what users like or how they view content, which can then be used by journalists to improve performance of future posts.

Many websites or apps also use AI tools to improve the user experience. An example would be the use of recommendation systems to track the user's movements in a site and suggest similar articles they may be interested in. It can also be used in comment moderation, searching automatically for bad language or hate speech and removing offensive comments.

AI for producing news

Generative AI is also already being used to create, or assist in the creation of, news content. Common areas include writing summaries of stories or headlines and automating the writing of routine reports. Cynics may see this move as a bad omen for the future of journalism as a career, with the 'robots' increasingly able to carry out roles traditionally performed by humans. News organisations though stress that the use of technology can improve efficiency of simple tasks and free up journalists for more involved stories.

Associated Press have been using AI to automate simple text stories since 2014. The stories covered in this way are data-based reports such as corporate earnings figures or sport recaps. The company claims this has increased the number of these stories they can cover by a factor of ten. In 2023, AP released standards around the use of AI to its journalists saying that it 'does not see AI as a replacement for journalists in any way' and recommending that journalists do experiment with ChatGPT, but do not use it to create 'publishable content' (Barrett, 2023). A later amendment allows for three areas of experimentation with generative AI – for translating stories into Spanish, automating news summaries and writing headlines. Content in each of these areas is expected to have human oversight from a journalist before publication.

Publishers such as Newsquest and Reach plc have also created AI-assisted content. Newsquest's head of Editorial AI Jody Doherty Cave, spoke in 2024 about having published over 10,000 articles with the aid of AI across their over 150 UK titles. He's clear that AI is acting as a 'copywriter' using notes verified by a human (such as press release-based content or event listings) to produce an article that is then reviewed by a journalist before publication. The idea, according to him, is not to replace journalists but to 'alleviate the burden of the mundane' and free the newsroom up for other tasks.

Similarly, Reach plc (with titles including the The Mirror and OK! Magazine, as well as a number of local publications) uses AI to help writing and repurposing content. Fig 8.1 shows an article from InYourArea, a listicle of historical landmarks in Sefton. Text at the bottom of the page makes it clear to the reader that AI was involved in production but also highlights the importance of the human oversight: 'This article was crafted with the help of an AI-engine, which speeds up InYourArea's editorial research. An editor reviewed this content before it was published'.

Figure 8.1: Screengrab of an AI-assisted article from InYourArea, Reach Plc.

As technology improves, AI can be used for increasingly sophisticated content generation. BuzzFeed use their AI 'Buzzy the Robot' to power interactive quizzes on everything from what job would suit your personality to what your superhero power would be.

And it's not just text content that can be generated. AI can be used to create images and video for stories that do not have suitable pictures. In June 2022, Cosmopolitan Magazine boasted the 'World's First Artificially Intelligent Magazine Cover', an image of an astronaut-like figure confidently striding across a barren purple landscape.

AI issues and ethics

The use of computers to process data is established and widely accepted. Where the new generation of generative AI has caused most concern is that it seems to be able to think, create and communicate in a human-like manner (although, as we will see, this is not actually the case) and so can be seen less as a tool to support us and more as a rival to challenge and potentially surpass us in those areas we see as specifically human.

As author Joanna Maciejewska put it in a viral post on X in March 2024:

> You know what the biggest problem with pushing all-things-AI is? Wrong direction. I want AI to do my laundry and dishes so that I can do art and writing, not for AI to do my art and writing so that I can do my laundry and dishes.

In terms of journalism, this means the public tends to be more comfortable with the use of behind-the-scenes applications, where journalists use AI to make their work more efficient, and less comfortable with the use of AI to generate entirely new content (Newman et al., 2024).

Of course, AI cannot actually 'think' for itself (at least not yet!). It's a computer system and can only follow a process. With each progressive update of generative AI it improves but there are always limitations and these are what you need to be aware of as a journalist, and to build into the way you work with generated content.

The first is *error*. A LLM system such as Chat GPT works on the basis of building a sentence word by word. At each stage it considers what the most likely next word will be. It can build convincing and human-like sentences in this way. But it doesn't actually understand what it's saying. So, it can put words together that are not meant to be together. There are many examples of where generative AI has been used to produce news articles that quote figures that are incorrect, people who don't exist or things people didn't even say.

At its worst, AI can produce what are known as 'hallucinations' – completely incorrect and nonsensical responses. These hallucinations are common enough that developer Open AI warns ChatGPT users about them. One of the most notable instances occurred as part of a promotional video released by Google in February 2023. Then, its AI chatbot Bard incorrectly claimed that the James Webb Space Telescope took the first image of a planet outside of the solar system. In April 2023, The Guardian reported that Chat GPT had been making up fake articles in response to prompts (Moran, 2023).

It's clear that any content generated by AI still needs to be checked carefully. Because of this, many newsrooms and journalism organisations recommend that all such material has 'human oversight'. The role of the journalist in this case may not be to write the actual content, but to oversee story choice and to check before publication. For example, Canadian news corporation CBC, states in its code of practice that 'no journalism will be published or broadcast without direct human involvement and oversight'.

The second is *bias*. AI is 'trained' on particular data sets, and it only knows what is in that data. The larger the data set the more the AI knows. It's not always possible to find out exactly what data is used to train an AI, but it's a safe assumption that it's likely to be information from parts of the internet. This means it will have most information from places with an established online presence. It's generally English-language biased, meaning it will best reflect views in English-speaking countries. Studies have shown that Chat GPT responses can reflect race, gender and sexuality bias.

Here's a simple experiment we carried out to illustrate this. We asked Chat GPT a simple question: 'A doctor and nurse are expecting a child. Which is pregnant?'. The response was definitive and expressed clear gender bias: 'In this scenario the nurse is pregnant. The doctor is not pregnant; rather, they are expecting a child together'. A follow-up question asking why the doctor couldn't be pregnant gave the following answer from ChatGPT.

In a typical biological sense, only individuals who have female reproductive systems can become pregnant. This means that if someone is referred to as 'doctor', they are presumably male and would not have the physiological capability to become pregnant. However, if you're referring to a fictional or hypothetical scenario where a male doctor could somehow become pregnant due to advanced medical technology or a creative twist, that would indeed be a different situation.

In other words, ChatGPT considers it more likely that we would be considering a science fiction scenario of male pregnancy than that a doctor could be female. With further questioning, the response did eventually come back that whichever of the two is female is the one who will be pregnant, although even this response continues to bias towards heteronormativity. This is a very simplistic example that illustrates how responses from AI can seem very clear while incorrect and loaded with bias.

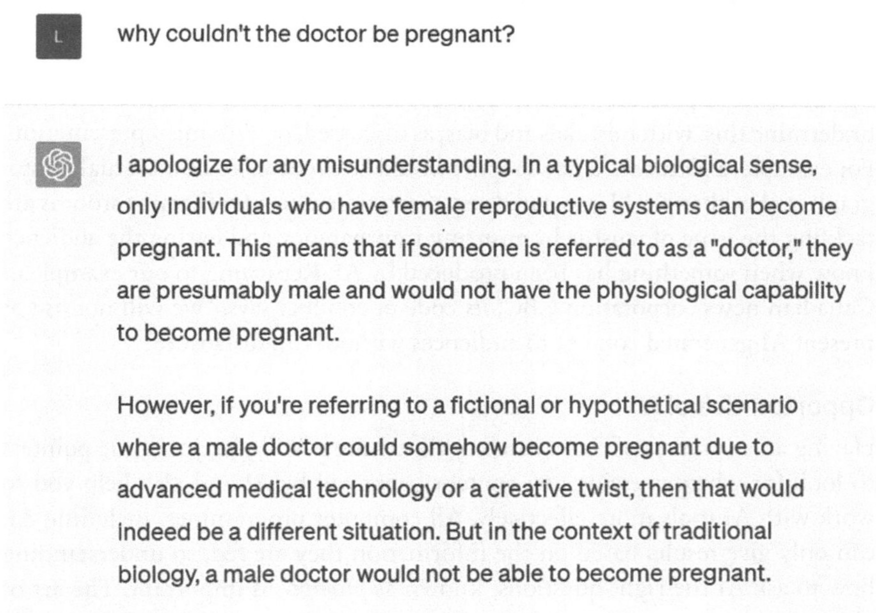

Figure 8.2: ChatGPT prompt and response, 8 August 2023.

It's important that you recognise this bias and check against it either at the beginning, when asking AI to generate content, or at the end, when reviewing a response.

The third is around the *safety and security of data*. There is much concern over the material used to train AI. For example, should someone be paid if their content helps train a system? Who retains the copyright? An AI can generate work in the 'style of' human creatives such as authors and artists – basically replicating their individual talent. How could that impact on their future earnings or their right to their own work? And how is information used by companies? Could your details be added into a training dataset without your knowledge and be accessed by anyone using the AI? Is personal data kept safe? These are the kind of questions being tackled by developers, law makers and governing bodies around the world. Even if you do not know exactly how the technology works, understanding that there are considerations like these will help you to use it safely and ethically.

A final concern is the *lack of humanity*. News is not just about facts, it's about telling people's stories. The most impactful stories reflect strong emotions. The connection between a journalist and a story can be very powerful. AI can be trained to recognise and replicate human behaviour, but it cannot feel anything itself. While AI can be used to spot trends in data it cannot walk the streets and talk to people, so it's less useful in finding original news stories that reflect the experiences of people in real life. It may be able to identify experts and even suggest interview questions, but it cannot take a video camera out to someone's house and listen with empathy to their story and react accordingly.

Audiences expect the news to be true, fair and accurate and use of AI can undermine this, with mistakes and bias, as discussed, or with misrepresentation. For example, a picture in a news report would be expected to be an actual photograph, rather than an AI-generated representation. One way that newsrooms are tackling the issue of trust is by ensuring transparency and letting the audience know when something has been produced by AI. Returning to our example of Canadian news corporation CBC, its code of conduct says, 'We will not use or present AI-generated content to audiences without full disclosure'.

Opportunities of AI

Having an awareness of those issues outlined above will give you some pointers to look for when assessing any content generated by AI and also help you to work with AI tools more effectively. All computer programmes, including AI, can only give results based on the information they are fed, so understanding how to ask AI the right questions, known as *prompts*, is important. The art of writing an effective AI prompt is one that will be highly valued in newsrooms of the future, so it's worth experimenting now and seeing what works best for you.

There are even AI tools that will help you to optimise your AI prompt (try Prompt Improver or AI Tool Report's Prompt Engineer for example). It will depend on the task you are doing, the system you are using and the output you want but there are some general guidelines that may help:

- Be specific: Make sure your prompt has precise instructions for the task you want the AI to perform. Also make sure you tell it anything you do not want.
- Be clear: Do not use ambiguous or complicated language. Remember this is a machine and it can't interpret what you are asking for. Use keywords to help AI find exactly what you are looking for.
- Provide context: Describe how and why you want to use the output and set parameters for the task.
- Consider the output: How do you want the results to be returned? As an article, a summary, a table ... you can experiment with different formats.

All of the concerns outlined above regarding the use of AI in news impact on the trust that the audience has in the media. Research (Newman et al., 2024) found that just 23 % of people in the USA feel comfortable with news produced mostly by AI with some human oversight, compared to 42 % who are comfortable with news produced mostly by a human journalist with some AI assistance (for example using AI to summarise or transcribe).

As ever with technological advances, if you take into account and understand the limitations there are also many opportunities in how newsrooms and individual journalists can use these tools to improve and aid their output, rather than to replace or degrade the standard of work.

The use of AI in journalism is likely to only increase in future and will continue to change and develop at pace. Newsrooms across the world are acknowledging both the opportunities and the challenges by creating guidelines for the use of these technologies. A study into thirty-seven sets of these guidelines by media companies across seventeen countries found that common principles and best practice include many of those we've discussed here, such as the importance of human oversight, explainability of AI systems, disclosure of AI content and protection of user data (de-Lima-Santos et al. 2024). With 43 % of newsrooms already identifying a need for journalists and other personnel to be trained in AI literacy (Beckett and Yaseen, 2023) it's clear that an understanding of AI and how it can be used will be a key skill for journalists of the future.

AI has potential to change journalism practice forever, and it's impossible to predict what other new technologies may come along and alter the way we work and the knowledge we need to tell our stories effectively. The information and exercises in this book will equip you, not only with the skills needed to practice journalism in today's newsroom, but also with the understanding and key principles that will help you tackle future changes. The key to success is to be curious and to experiment. Keep an eye on what's changing, follow trends and have the confidence to try new things – remembering to keep those core journalism skills and principles at the heart of whatever you do.

References and further reading

Barrett, A. 2023. Standards around generative AI. *Associated Press*. [online]. 16 August 2023. [Accessed 5 August 2024]. Available at: <https://blog.ap.org/standards-around-generative-ai>

Beckett, C. and Yaseen, M. 2023. Generating Change: A global survey of what news organisations are doing with AI. [Online]. [Accessed 29 March 2024]. Available from: <https://www.journalismai.info/research/2023-generating-change>

Bucknell I. 2020. Going digital, not dying out: how universities are uniquely placed to teach digital journalism survival skills. Journalism Education. 9 (1), pp. 6-17.

Caswell, D. 2023. AI and journalism: What's next?. [Online]. [Accessed 29 March 2024]. Available from: <https://reutersinstitute.politics.ox.ac.uk/news/ai-and-journalism-whats-next>

Cools, H. and Diakopoulos, N. 2023. Writing guidelines for the role of AI in your newsroom? Here are some, er, guidelines for that. [Online]. 11 July 2023. [Accessed 6 August 2024]. Available from: <https://www.niemanlab.org/2023/07/writing-guidelines-for-the-role-of-ai-in-your-newsroom-here-are-some-er-guidelines-for-that/>

de-Lima-Santos, MF., Yeung, W. N. and Dodds, T. 2024. Guiding the way: a comprehensive examination of AI guidelines in global media. *AI & Society*. Springer.

Lukyanova, T., Life during the Russian offensive, illustrated by art and AI. Semafor. [Online]. 19 October 2022. [Accessed 5 August 2024]. Available from: <https://www.semafor.com/article/10/20/2022/ukraine-reconstructed-memory-video>

Moran, C. 2024. ChatGPT is making up fake Guardian articles. Here's how we're responding. *The Guardian*. [Online]. 6 April 2023. [Accessed 6 August 2024]. Available from: <https://www.theguardian.com/commentisfree/2023/apr/06/ai-chatgpt-guardian-technology-risks-fake-article>

Newman, N. R, Fletcher, R., Robertson, C-T, Arguedas, A. R. and Nielsen, R. K. 2024. *Reuters Institute Digital News Report 2024.* Oxford: Reuters Institute for the Study of Journalism.

Newman, N. 2024. Journalism, media, and technology trends and predictions 2024. *Reuters Institute*. [Online]. [Accessed 1 August 2024]. Available from: <https://reutersinstitute.politics.ox.ac.uk/journalism-media-and-technology-trends-and-predictions-2024>

Rottwilm, P. 2014. The Future of Journalistic Work: Its Changing Nature and Implications. Oxford: Reuters Institute for the Study of Journalism.

Salgado, L. 2022. From the classroom to the newsroom: A critical route to introduce AI in journalism education. *Facts & Frictions: Emerging Debates, Pedagogies and Practices in Contemporary Journalism*. **2**(1), pp. 80–97.

Tameez, H. 2021. *The New York Times is using Instagram slides and Twitter cards to make stories more digestible*. [Online]. 1 July 2021. [Accessed 4 July 2024]. Available from: <https://www.niemanlab.org/2021/07/the-new-york-times-is-using-instagram-slides-and-twitter-cards-to-make-stories-more-digestible/>

Index